Sucrochemistry

Sucrochemistry

John L. Hickson, EDITOR

Hickson Consulting Services

A symposium sponsored by
The International Sugar Research
Foundation, Inc., and by the
Division of Carbohydrate Chemistry
at the 172nd Meeting of the
American Chemical Society,
San Francisco, Calif.,
Aug. 31–Sept. 2, 1976.

ACS SYMPOSIUM SERIES 41

AMERICAN CHEMICAL SOCIETY
WASHINGTON, D. C.   1977

Library of Congress CIP Data

Sucrochemistry.
(ACS symposium series; 41 ISSN 0097-6156)

Includes bibliographical references and index.

1. Sugar—Congresses.
I. Hickson, John L., 1916– . II. International Sugar
Research Foundation. III. American Chemical Society.
Division of Carbohydrate Chemistry. IV. Series: Ameri-
can Chemical Society. ACS symposium series; 41.

QD320.S8   547'.781   77-1296
ISBN 0-8412-0290-7   ACSMC8 41 1–381

This volume is dedicated

to the memory of:

Ody H. Lamborn, gentleman and persuasive thinker, who, in 1941, conceiving of the opportunities in research on the utilization of sugar for the good of mankind, persuaded the sugar industry to organize the Sugar Research Foundation, Inc.

and as an accolade to:

Henry B. Hass, chemical ideator extraordinaire, who, in 1952, ignited the spark of sucrochemistry and generated the term to identify it.

ACS Symposium Series

Robert F. Gould, *Editor*

FOREWORD

The ACS Symposium Series was founded in 1974 to provide a medium for publishing symposia quickly in book form. The format of the Series parallels that of the continuing Advances in Chemistry Series except that in order to save time the papers are not typeset but are reproduced as they are submitted by the authors in camera-ready form. As a further means of saving time, the papers are not edited or reviewed except by the symposium chairman, who becomes editor of the book. Papers published in the ACS Symposium Series are original contributions not published elsewhere in whole or major part and include reports of research as well as reviews since symposia may embrace both types of presentation.

CONTENTS

PREFACE

The drive for relevance by a commodity industry can be recognized through many guises. One example is the age-old sugar industry. A capsulated version of its long and interesting history may be in order. The original discovery of the sweetness of the juice of the sugar cane is lost in prehistory, but wild sugar canes are still indigenous to the upper regions of those major rivers of India. A custom native to that region, the chewing of a "honey grass," is noted by the historians of Alexander the Great. Over the years this evolved into a lucrative sugar trade throughout the Mediterranean community. This was one of the rich prizes won by the Venetians as they wrested preeminence in the Inland Sea. The Venetians improved the commodity with a refining process, and cultivation of sugar cane spread Westward along both shores of the Mediterranean through the 13th–15th centuries. When, in 1493, Columbus was sent back to Hispaniola with colonists for the New Spain, they introduced the culture of sugar cane to their New World. Sugar beets, a thriving agribusiness in Europe since the Napoleonic wars, were introduced into the U.S. toward the middle of the 19th century.

Thereafter, advancing technologies in agronomics, unit chemical engineering processes, and transportation transformed sugar from an expensive delicacy to a household and manufacturing staple. This availability, coupled with the rising standard of living in the latter part of the 19th and the 20th centuries, erased former economic barriers to its use. The populace of the United States and much of the North Atlantic community responded in the years following the First World War by sating their appetites with 30–50 kg/caput/annum, at which level it has remained now well into the 7th decade of this 20th century. Thus the markets for the ancient sugar industry have indeed matured.

This maturation for the sugar markets of the United States was comprehended by some sugar men in the early years of the Second World War. Challenged by the late Ody H. Lamborn in 1943, the leaders of the U.S. sugar industry elected to grasp an opportunity for improvement by organizing the Sugar Research Foundation. The goal established then, and still viable for this 33-year-old organization, is to defend and extend the utilization of the commodity—sugar.

Then, as now, at least 98% of the sugar sold in the United States was consumed by humans, most of it processed in the food business.

xiii

Yet there was a feeling of optimism that, if enough could be learned about the chemical peculiarities of this substance, ways could be devised to convert it into many of the material things required by a modern society. In such a way sugar could be relevant to the present age in a role secondary to its gustatory and nutritive values.

Only the more foresighted of the pundits of that era were forecasting an end to the age of liquid and gaseous fossil fuels, a reality that would be made painfully evident to us. Hardly anyone could foresee a forthcoming age in which the focus of such a large portion of the materials industries would be turned toward renewable resources to supply the voracious appetites of the industries based on the organic compounds of carbon. Yet the research ethic, woven into the framework of the Foundation by O. H. Lamborn, and the continuing support of research by the leaders of the sugar industry have created a refrain which is consonant with what turns out to be the great theme of the 1970s—the substitution of renewable resources for those failing fossil carbon resources.

The goal itself has not yet been achieved, for the sugar molecule is obstreperous. Sugar is an abundant resource of nearly matchless, molecular homogeneity, available at prices seemingly reasonable. Yet the prophet of utilization who has dared to roll up his sleeves and attempt sugar reactions in his laboratory, more often than not, has met stark frustration. The small differences in the energies of activation among the several classes of hydroxyl groups in sucrose, for all but the most sophisticated of chemists, have produced intractable and almost unresolvable mixtures of positional isomers with ranges of degrees of substitution. Most investigators have retreated to "safer and tamer" starting materials.

In 1952, however, the Board of Directors enticed a demonstrated chemical innovator to accept the presidency of the Sugar Research Foundation. Henry B. Hass had created quite a stir in the chemical community by taming the vapor phase reactions of hydrocarbons to produce halogen and nitro derivatives with intriguing potentials. Dr. Hass brought to the research program of the Foundation new ideas and boundless enthusiasm. He charged research groups with seemingly irresistible drives to tackle sucrochemical problems, and his infectious enthusiasm fostered a continuing flow of "patient money." In the 24 years since Dr. Hass initiated the program, enough of a harvest has accrued from the seed he planted to justify this symposium. As assistant to Dr. Hass for seven of those years and subsequently for eleven years as Director of Research for the Foundation, your editor has found excitement and reward in the challenge of sucrochemistry.

In the gathering experience of the midcentury, the sugar men of many nations likewise recognized that, as in the United States, their markets had attained a stable maturity. Many accepted the rationale of the premises upon which the Sugar Research Foundation was structured. Sugar companies in Canada, Europe, Africa, and Australia joined in the support of this research program, thus establishing centers of sucrochemical research in many lands. Cognizance of this international character of the program is evidenced by the fact that 14 of 27 papers in this symposium are written by non-American authors.

In 1968 the Foundation likewise acknowledged this broadening of its support by redesignating itself as International Sugar Research Foundation and strengthening its commitment to the encouragement of research on the uses of sugar. One of the vehicles employed by ISRF is the fostering of symposia on a variety of subjects. In this role ISRF welcomed the invitation from Milton S. Feather (University of Missouri), then Chairman-Elect of the Division of Carbohydrate Chemistry of the ACS, to set up a symposium on sucrochemistry. With the strong endorsement of Gilles E. Sarault, President, and with a significant commitment from the Board of Directors of ISRF, the writer was assigned the task of creating the symposium.

In the structuring of this program, no attempt has been made to be encyclopedic; rather the aim has been to display a sampling of the work by some of the star performers, selected to illustrate how some of the difficulties have been surmounted and to recount some of the rewarding achievements in this quarter century of exploration. In the first section the concepts and evolution of an intriguing, fundamental chemistry of the sucrose molecule are explored. Three sections are devoted to illustrating some of the industrial applications of sucrochemistry: in surfactants, surface coatings, urethane plastics, and fermentation processes. The symposium closes with discussions of the business and economic forecasts for sucrochemistry.

It has been a distinct privilege and pleasure, both personally and professionally, to draw together these 27 papers from outstanding contributors representing academic and industrial leaders from seven countries. These contributions are no less significant than they are self-evident.

Yet, other major contributions to the achievement of the success of the symposium deserve to be recognized. First, the International Sugar Research Foundation contributed generous financial backing and staff time. The typing of the manuscripts for the printing was accomplished by the diligent, dedicated, cheerful, and thoughtful labors of Ms. Elizabeth Dodds. Finally, and most significantly, an editor could not wish for greater cooperation than the time, energy, initiative, and application of intelligent editorial skills given by Stephanie S. Hillebrand (Information

Officer at the Foundation), who has served so capably as the assistant editor on the manuscript.

Working with such a team of contributors, backers, and production staff, it has been most rewarding to have created this gateway on the path to relevance in an age of sucrochemistry.

5915 Bradley Blvd.   John L. Hickson, Consultant
Bethesda, Md. 20014
December 15, 1976

Concepts and Basic Discoveries

Introduction

WHITNEY NEWTON II

Holly Sugar Corp., P.O. Box 1052, Denver, Colo. 80901

Many interesting papers have been prepared for presentation at this meeting which, I believe, will help bring us all up to date on new developments pertaining to reactions of sucrose, analytical methods used to measure the results, and commercial applications of sucrose chemicals.

More interest will be shown in compounds that can be made from raw materials that are renewable. Thus, as the price of crude oil continues to climb, more products from agriculture will be required as the raw materials for the chemical industry. Sucrose, molasses and bagasse will find a real place in this raw materials market.

At this time reactions of sucrose are known that will produce many of the chemicals of industry, and more are certain to develop since sucrose can be grown in many different climates with yields in calories per acre exceeding most agricultural crops. It is therefore reasonable to believe sucrose will fill a real gap left by the world shortage of crude oil.

I think it is very fitting that Henry Hass is the first speaker of this Symposium. Dr. Hass has probably contributed more to stimulate interest in sucrochemistry than any other single individual. Perhaps his greatest contribution is the interest he has shown in developing carbohydrate chemists. In addition, as president of the Sugar Research Foundation, he was instrumental in funding a large portion of the research that has been done in the sucrochemical field, and which will be reported on during the next few days.

1

The Concept of Sucrochemistry

H. B. HASS

Consultant, 95 Fernwood Rd., Summit, N.J. 07901

By sucrochemistry is meant the branch of science and technology whose objective is to bring added markets to sucrose and its byproducts by their chemical utilization. It is thus one sector of the much broader area called chemurgy. The chemurgic movement coined a neologism based upon the concept of putting chemistry to work to solve the problem of surpluses of farm crops. Scientists and crossword puzzle addicts will recognize the familiar term erg, a unit of work or energy, here spelled with a "u" to avoid problems of pronunciation.

At present, crop surpluses have been largely wiped out by a burgeoning of world population, but the approaching depletion of our petroleum resources and the realization that even coal, oil shale and tar sands are not infinite in amount have emphasized the importance of the utilization of renewable resources such as forest products and annual crops. As we burn more and more fuels to carbon dioxide we can expect, in the long run, a more rapid growth of vegetation. The last time that we had plenty of carbon dioxide in the air, during the carboniferous era, the ferns grew 200 feet high. Thus the long view of industrial organic chemistry must inevitably foresee a continually decreased dependence upon fossil fuels. Sucrochemistry must inevitably be part of the wave of the future. Nobody can now know how much of that wave will be sucrochemical in nature, but the uniquely high productivity of sugar cane and sugar beet suggests that they will serve as very important sources of chemicals.

*From 1952-1960, Dr. Hass was President and Director of Research, The Sugar Research Foundation, Inc. New York which, in 1968, became the International Sugar Research Foundation, Inc., Bethesda, Maryland, U.S.A.

It is well known to sugar chemists, though not to
chemists in general, that sucrose is the pure organic
chemical produced in largest amounts, world-wide.
Ethylene has been creeping up on sucrose but is not
yet its equal in annual tonnage. Before our efforts
at Sugar Research Foundation, sucrose and its byproducts had been used to make such products as ethanol,
butanol and acetone by fermentations, Celotex, wall-
board and paper from sugarcane bagasse, which also
served as a raw material for furfural. The growth of
the petrochemical industry was rapidly superseding the
fermentation processes so it seemed that our most successful approach would be not in the direction of making simpler organic molecules, but to utilize the
unique chemical structure of sucrose to build molecules
which could not, as a practical matter, be duplicated
from petrochemical sources.

Since the sucrose molecule contains three primary
and five secondary hydroxyl groups, these were the logical points of attack. Sucrose had previously been
converted to the octaacetate which, being intensely
bitter, found a minor use as a denaturant for ethanol.
It seemed to me that putting one or two lipophilic
groups on sucrose would be likely to generate a surfactant because of the highly hydrophilic nature of
sucrose. As then Director of Research of the
Sugar Research Foundation, I consulted Foster D. Snell
on this idea because he was very knowledgeable about
surfactants.

Foster Snell liked the idea of sugar-based surfactants so I wrote out twenty ways in which it seemed
that this objective could be realized. One of the
problems was to find a suitable solvent for sucrose
and something very lipophilic. Dr. Lloyd Osipow,
then of the Snell organization, tried dimethylformamide
and it worked. By transesterification under mildly
alkaline conditions a good yield of monoester could be
obtained and we were on our way. The equation is
deceptively simple.

$$C_{12}H_{22}O_{11} + RCO_2CH_3 = RCO_2C_{12}H_{21}O_{10} + CH_3OH.$$

Unfortunately, the bad effects of traces of water, the
difficulty of completely drying sucrose crystals, which
tend to trap mother liquor from the syrup undergoing
crystalization, the problem of removing the last traces
of dimethylformamide from the product and the reser-
vations of the U.S. Food and Drug Administration (FDA)
over the possible toxicities of residues were responsible for much trouble in the development stage.

Then there occurred the era of the concept of zero tolerance for toxic materials, which is now pretty well behind us and this is fortunate, for it meant that, as analytical methods for detecting traces improved, so did unacceptability of products. Thus, the manufacturer was always shooting at a moving target. Further, the concept of zero tolerance is based upon the assumption that if a large amount of a material is toxic, one tenth as much will be about one tenth as toxic. This is not only unscientific but is contrary to common experience. A good full meal makes you feel great; twice as much gives you a stomach ache. One drink makes people talk more vivaciously; ten drinks put you to sleep. Many of the essential nutrients are toxic in large amounts. The list includes common salt, fluoride ion, cobalt and vitamin D. The essential elements selenium and iodine are both toxic if you ingest too much. The reductio ad absurdum of the idea of zero tolerance of poisons came up when vitamin B_{12} was found to contain cyanide; still it is an essential nutrient.

The net effect of this questionable attitude thrust upon FDA was that sucrose esters went into commercial production in Japan rather than in the land of their discovery. In Japan they go into cake mixes as emulsifying agents, and so far they have hurt no one. The process of digestion splits the sucrose ester into fatty acid and invert sugar, compounds that are normally present in the body. More recently I have seen reports that the French are using sucrose esters in animal feeds. The results of experiments synthesizing sucrose esters from acid anhydrides, to be reported later in this symposium, are, naturally, of great interest to me.

Meanwhile I got the idea that superior drying oils might be produced if we crowded as many drying oil fatty acid molecules as possible into a sucrose molecule. Since this was almost entirely a development problem, we turned it over to Professors Edward G. Bobalek and T.J. Walsh at Case Institute of Technology. They found that sucrose esters have advantages over ordinary drying oils in that they form harder, glossier films which adhere to metals better and wrinkle less on drying. These effects seem related to the octafunctionality of sucrose versus the trifunctionality of glycerol. Thus a hexa- or hepta ester of sucrose has a higher molecular weight than its glycerol or pentaerythritol analog containing the same fatty acids and thus has many more possibilities of joining with adjacent molecules in the normal drying processes.

We restudied the ammoniation of beet pulp which had been reported by one of our member companies but was not being used commercially. The chemistry involved is the ammonolysis of methyl ester groups in the beet pulp to liberate methanol and form amide structures. These amides are attacked by enzymes produced by microorganisms in the rumen of cattle and sheep to form protein which becomes available to the animals. In research at the National Institute for Research in Dairying, in England, the product was found to be equivalent to peanut meal as a source of protein for growing calves, but only after an induction period during which the microorganisms of the rumen were adjusting their relative numbers to the new feed.

We tried to replace the OH groups of sucrose with NH_2 groups by the simultaneous action of ammonia and hydrogen. This works well with simple alcohols but the sucrose molecule goes all to pieces under these conditions and the principal product, isolated by Professor Philip Skell at Pennsylvania State University was 2-methylpiperazine. Efforts to improve the yield sufficiently to make this an efficient source of diamines for nylon-type polymers have not so far been successful.

An expired U.S. Patent on the use of sucrose along with phenol and formaldehyde to make condensation resins was investigated with inconclusive results. The hope was that sucrose, then very inexpensive, could be built into Bakelite-type polymers without too much sacrifice of properties. Sucrose-urea resins were also investigated with poor results.

We started work on sucrose acetals with Professor E.J. Bourne at Royal Holloway College in England and these compounds were reported for the first time. It is of interest to me that Dr. Riaz Khan is continuing this work at Tate and Lyle, Ltd., and will present his results later in this symposium.

We started a project with Professor Harry Szmant at University of Oriente, Cuba, on replacing sucrose hydroxyl groups with chlorine, but the Castro revolution there interfered. I am pleased that related work is continuing in this field.

This brings us to the end of this part of the story. In the words of the poet John McRae,
"To you from falling hands we throw the torch
Be yours to hold it high!"

Abstract

The use of sucrose and its byproducts as raw mat-

erials for chemical manufacture is much older than my
efforts in this direction.   These included Celotex
sound-absorbent wall-board from sugarcane bagasse,
typewriter paper from bagasse, ethanol, butanol and
acetone by fermentation from molasses, though these
fermentations were being superseded by petrochemical
processes.

"Sucrochemistry" covers a branch of science and
technology whose objective is to bring added markets
to sucrose and its byproducts by chemical utilization.
It is thus a part of the broad area of chemurgy and is
based upon the concept that, although the capacity of
the human stomach is limited, mankind's desire for
manufactured goods seems not to be.   The pure organic
chemical manufactured worldwide in the largest tonnage
is sucrose, but it had been relatively little studied
compared to benzene, for example.   We restudied the
ammoniation of dried beet pulp to make a protein-equiv-
alent feed for ruminants and we also ammoniated bagasse
and molasses for the same purpose.   We made sucrose
mono- and difatty acid esters as edible surfactants,
subjected sucrose to reductive aminolysis, used sucrose
in the formation of phenolformaldehyde resins, and
made drying oils from sucrose and unsaturated fatty
acids.

Biographic Notes

Henry B. Hass, Ph.D. (and 6 honorary degrees).
Retired educator and consultant to the chem. ind.
Educated at Ohio Wesleyan Univ. and Ohio State Univ.
For 21 years on the staff and 12 years as Head, Chem.
Dept. Purdue Univ.   Industrial experience: General
Aniline & Film Corp., then, for 8 years, Pres. and
Dir. Res. Sugar Research Foundation, Inc.; then Dir.
of Chem. Res., M. W. Kellogg Co.   Some 200 papers and
150 U.S. patents.   A consultant in industrial chemis-
try.   95 Fernwood Drive, Summit, N.J. 07901 U.S.A.

Selective Substitution of Hydroxyl Groups in Sucrose

LESLIE HOUGH

Chemistry Department, Queen Elizabeth College, Camp Hill Rd.,
London W8 7AH, England

The chemistry of sucrose (A), often termed 'Sucro-chemistry', has played a special role in our studies of carbohydrates in view of its ubiquity and great impor-tance in commerce. Our interest has been concentrated upon stereoselective chemical reactions, in the main by replacement of specific hydroxyl groups by other functional groups. The considerable progress made in this field, despite the complexities associated with the chemistry of this unique molecule, have been due to the application of mass spectrometry, [1]H n.m.r. and [13]C n.m.r. to the structure determination of the pro-ducts obtained by simple but effective chromatographic control and, in the case of complex mixtures of pro-ducts, through purification by column chromatography. A profile of chemical reactivity has emerged such that wider applications in sugar and associated technologies can now be expected as the fossilized materials become increasingly expensive. Sucrose and its derivatives show stereoselectivity in many of their reactions, which is fortunate, since a large number of partially substituted derivatives are theoretically possible. (See Table I).

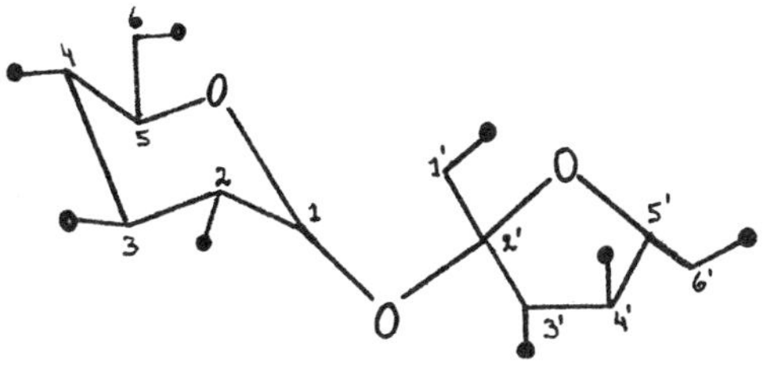

(• = OH)   **A**

Table I

No. of Isomers of Sucrose Derivatives

MONO	8	PENTA	56
DI	28	HEXA	28
TRI	56	HEPTA	8
TETRA	70	OCTA	1

Thus, whilst only one octaderivative of sucrose can
exist, such as the octamesylate (B), there are 70 possible tetraderivatives and 56 alternatives in the case
of both tri- and penta-derivatives. Our studies (1,2)
have revealed that the eight hydroxyls of sucrose (A)
react selectively and four, including the three primary
groups, can be replaced to give pure 6,6'-di-, 4,6,6'-
tri- and 1',4,6,6'-tetrasubstituted products, dependent upon the reactants and the reaction conditions.
For example, when sucrose octamesylate (B) was treated
with nucleophiles, such as iodide, bromide, chloride,
azide, etc, in an aprotic solvent, such as N,N-dimethyl-
formamide (D.M.F.) or hexamethylphosphoric triamide

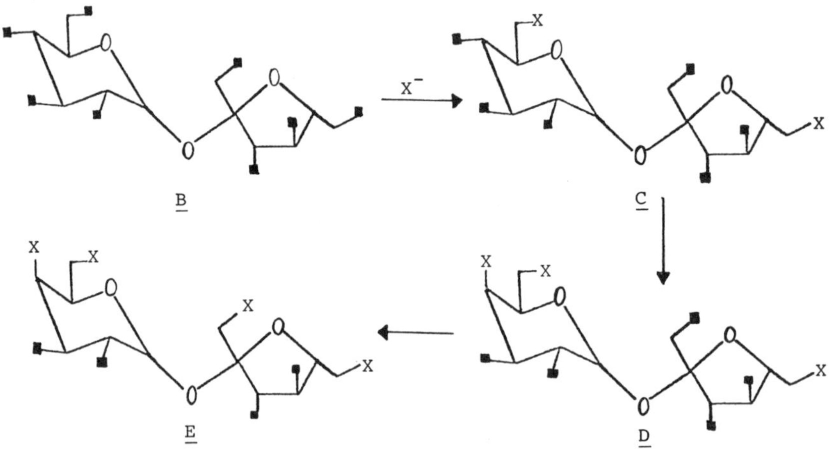

$(\blacksquare = O-SO_2CH_3)$

$(X = \text{Nucleophile})$

(H.M.P.T.), specific mesyloxy substituents underwent
nucleophilic substitution to give 6,6'-disubstituted
(C), 4,6,6'-trisubstituted (D) and 1',4,6,6'-tetra-
substituted (E) products that could readily be isolated
by careful control of the reaction conditions and subsequent purification by column chromatography. It was
established that the order of replacement is 6∿6'>
4 >1' ≫ all other positions.

The observed stereoselectivity (1,2) is in accord
with the bimolecular transition state theory (3).
The substituents at the least hindered, primary 6- and
6'-positions reacting preferentially and the other primary substituent, of the more crowded neo-pentyl type
at the 1'-primary position, reacting more slowly than
the more favorable, but secondary, 4-position. In
accord with this mechanism, inversion of chirality is
always observed when nucleophilic substitution occurs
at a chiral centre, in this case at C-4. By this process, 4-sulphonate esters of sucrose are converted by
substitution and inversion of configuration at C-4 into
β-D-fructofuranosyl α-D-galactopyranoside (F)
('galacto'-sucrose) and its derivatives (D and E)
(4,5). It is of interest to note that galacto-sucrose
(F) is not sweet, due to the presence of an axial 4-
hydroxyl group (6) (see later, however).

F

By selective esterification with sulphonyl halides
partially substituted sulphonate esters of sucrose can
be prepared. For example, using limited quantities of
toluene-p-sulphonyl ('tosyl') chloride in pyridine and
subsequent purification by column chromatography, the
6,6'-ditosylate (G), 1'6,6'-tritosylate (H), and 1',
2,6,6'-tetratosylate (I) were obtained. Advantage
was subsequently taken of the increased selectivity of
bulkier sulphonyl halides (7), such as triisopropylben-
zene sulphonyl chloride ('tripsylchloride') and mesity-
lene sulphonyl chloride to prepare the corresponding
1',6,6'-trisulphonate esters directly and in good yield
(8,9), without the use of column chromatography.

A   G

I   H

(• =OH)

(*TsCl = Tosyl Chloride
  py = Pyridine)

Nucleophilic substitution of the sulphonyloxy substituents in these partially substituted sulphonate esters
of sucrose readily affords 6,6'-di- and 1',6,6'-tri-
substituted sucrose derivatives, such as the corresponding chlorides and azides, from which the deoxy- and
amino-derivatives,respectively, can be prepared.

   As an alternative, the selective de-esterification
of sucrose octaacetate (J) was examined. When a
chloroform solution of the latter (10) was passed
through a column of alumina, three heptaacetates were
obtained with free hydroxyls at C-6', C-4' and C-4
respectively (K,L and M) (10). Since acetyl groups
readily migrate, via an 4,6-orthoester intermediate,
from the 4- to the 6-position of the D-glucopyranosyl
ring, and presumably less readily from 4'- and 6'-,we
(10) suggested, with supporting evidence, that the
initial products were the two heptaacetates, with the
6-hydroxyl and the 6'-hydroxyl respectively free.
The final products probably resulted from acetyl migration.

   Selected hydroxyls in sucrose can be replaced by
chloride by exploiting reactions with either sulphuryl
chloride in pyridine (11) or mesyl chloride in D.M.F.
(12,13). The reaction with sulphuryl chloride involves
the initial formation of a chlorosulphate ester (N)
which then undergoes an intramolecular nucleophilic
substitution reaction with the insertion of chloride.

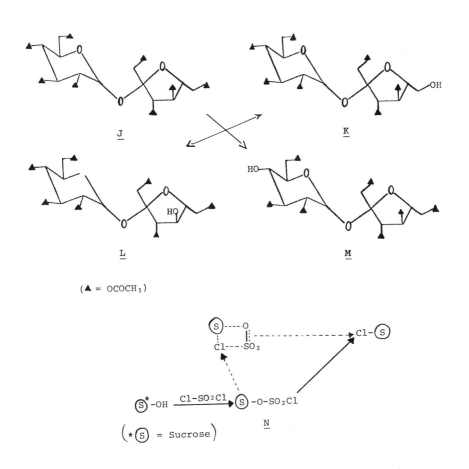

J

K

L

M

(▲ = OCOCH₃)

N

$\left(* \text{ⓈⓈ} = \text{Sucrose} \right)$

O

P

Under controlled conditions, sucrose reacts with
sulphuryl chloride in pyridine in a highly selective
manner to give either 6,6'-dichloro-6,6'dideoxysucrose
(O) or 4,6,6'-trichloro-4,6,6'-trideoxy-galacto-sucrose
(P) which can be obtained directly, after de-chlorosul-
phation, without resorting to preparative chromatography. (16) In these substitution reactions, the
least crowded, primary 6- and 6'-positions react preferentially and the favorable secondary 4-position
reacts faster than the more hindered but primary
1'-position. In a related study Parolis (14) has
also isolated the 4,6,6'-trichlorosucrose (P) directly
but as the crystalline 1',2,3,3',4'-pentachloro-sul-
phate. More extensive reaction of sucrose occurs with
sulphuryl chloride at higher temperatures to give three
4,6-dichloro-4,6-dideoxy-α-D-glucopyranosyl 2,3-sul-
phate derivatives (Q,R and S), differing in the modifications to the fructofuranosyl moiety (15). The chlorosucroses (O) and (P) are readily converted into anhydrides (16) and deoxyderivatives (17). Thus the
6,6'-dichloride (O) was readily transformed into the
3,6;3',6'-dianhydroderivative (T) when treated with
methanolic sodium methoxide, whereas, under similar
conditions, the 4,6,6'-trichloride (P) gave initially
the 3,6-anhydro-4,6'-dichloride (U), due to rate-en-
hancement by the axial 4-chloro group which moves to
the more favorable equatorial position ($^4C_1 \rightarrow {}^1C_4$)
followed by the formation of the final product, the
4-chloro-4-deoxy-3,6;3',6'-dianhydro derivative (16)
(V).

(• = OH)

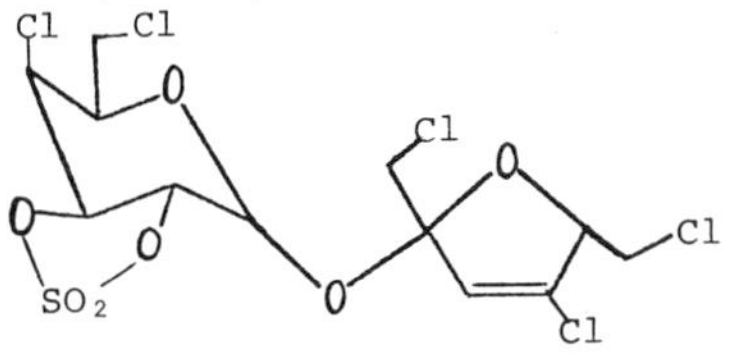

Q

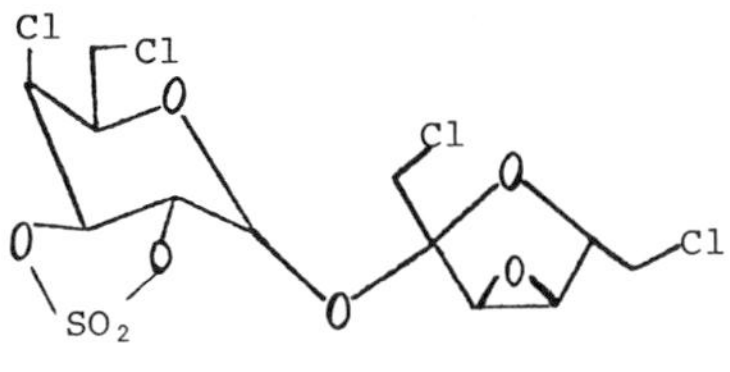

R

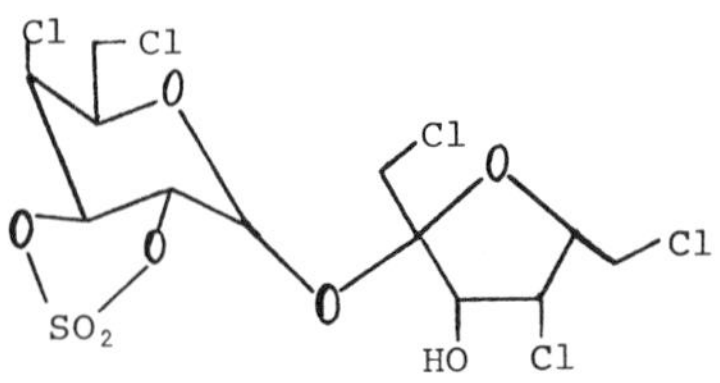

S

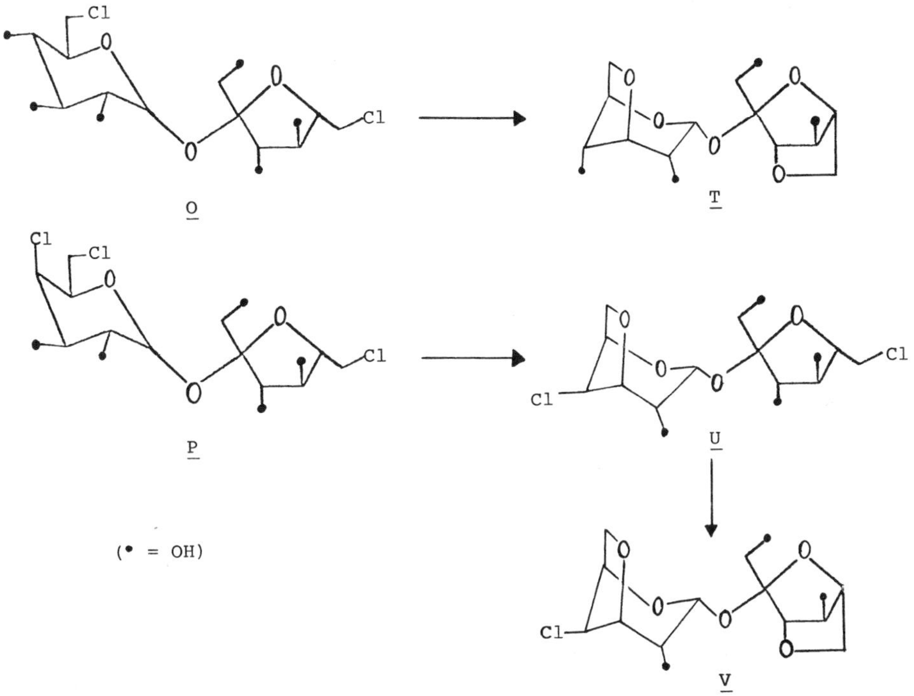

$\underline{O}$ $\underline{T}$

$\underline{P}$ $\underline{U}$

$(\bullet = OH)$

$\underline{V}$

$\underline{P}$ $\underline{W}$

$\underline{X}$ $\underline{Y}$

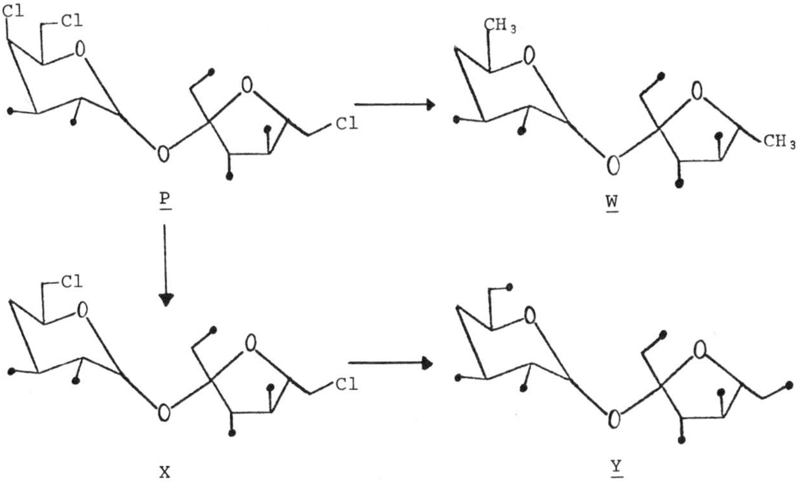

$(\bullet = OH)$

The 4,6,6'-trichloride (P) was converted into
4,6,6'-trideoxysucrose (W) by catalytic, reductive
dehalogenation in the presence of potassium hydroxide
(17). On the other hand, in the presence of triethylamine, dechlorination occurred exclusively at the secondary 4-position to give the 4-deoxy-6,6'-dichloride
(X) as expected from the results of Lawton, Wood,
Szarek and Jones (18). Nucleophilic replacement of the
6,6'-chloro-substituents of the latter (X) by benzoate
then gave 4-deoxy-sucrose (17) (Y). The isomeric 1',
6,6'-trichloro-1',6,6'-trideoxysucrose (AA) was con-
veniently prepared from 1',6,6'-trimesitylene sulpho-
nylsucrose (Z), and then converted as above into 1',
4';3,6;3',6'-trianhydrosucrose (BB) and 1',6,6'-tri-
deoxysucrose (CC) (19). Further chlorination of the

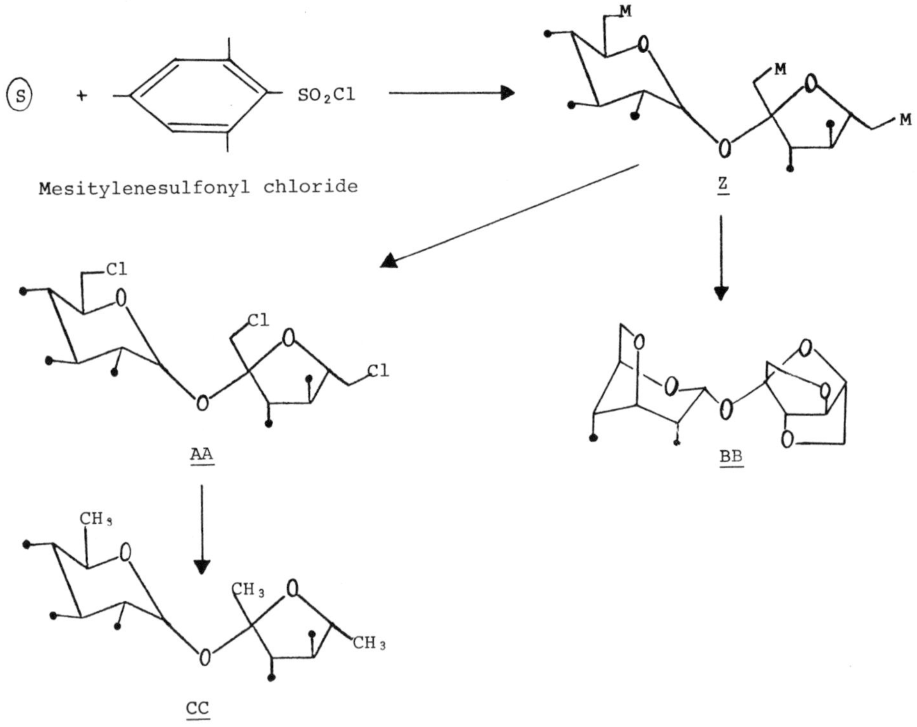

Mesitylenesulfonyl chloride   Z

AA   BB

CC

(S) = Sucrose)

(• = OH)

(M = Mesitylenesulfonyloxy)

1',6,6'-trichloride (AA) by reaction with sulphuryl
chloride in pyridine occurred at C-4 as predicted, to
give 1',4,6,6'-tetrachloro-1',4,6,6'-tetradeoxy-galac-
to-sucrose (DD). An objective in our chemical studies
on sucrose has been to enhance its natural sweetness,
and we have been considerably encouraged in this
direction by the surprising discovery that this tetra-
chloride (DD) is intensely sweet, comparable to saccharin but with a pleasant after-taste (20). The apparently contrary loss of sweetness in galacto-sucrose (F)
has been attributed to hydrogen bonding from the axial
4-hydroxyl group to the ring oxygen which cannot, of
course, occur at C-4 in the sweet 1',4,6,6'-tetrachlo-
ride derivative (DD). The considerable enhancement of
the sweetness of sucrose by selective substitution by
chloride, not observed hitherto in this disaccharide
nor in any other carbohydrate, is clearly of importance
not only in nutrition in the development of alternative
sweeteners to sucrose, but in relation to theories of
sweetness.

AA   DD

EE   FF

(• = OH)

Application of ^{13}C n.m.r. to the characterisation of derivatives of sucrose has revealed (21) that in addition to ^{1}H n.m.r. and mass spectrometry, it is a powerful aid since the position of replacement of hydroxyls by other substituents and their stereochemistry is easily recognised by the shifts. The resonances of C-1', C-6 and C-6', the primary carbons in sucrose, have been assigned (21) (Figure 1) and are distinctive in that they occur at high field. The signals for C-2, -3 and -5 of sucrose (A) have not so far been assigned unequivocally, but they were unaffected by the introduction of chlorine at C-6 whereas epimerisation at C-4 to give galacto-sucrose (F) results in upfield shifts of C-2, -3 and -5 (Figure 2). In an attempt to assign the signal for C-3, sucrose (A) has been oxidised to 3-keto-sucrose (DD) by Agrobacterium tumefaciens (22) and the product (EE) reduced with sodium borodeuteride, which showed the expected stereospecificity to give {(3-^{2}H)- α -D-allopyranosyl β -D-fructofuranoside (FF); 'allo'-sucrose} (23).

Acknowledgements

We are grateful to the International Sugar Research Foundation Inc. and Tate & Lyle Ltd. for their continued support and interest in these studies.

Abstract

In studying the fundamental chemistry of sucrose over the past two decades, we have discovered a profile of chemical reactivity that is unique to sucrose and devised procedures for the isolation and characterisation of the products. Thus precise control of the reactions of sucrose with sulphuryl chloride in pyridine can effect replacement of two, three or four hydroxyl groups in sucrose. By variation of the reaction conditions we can isolate 6,6'-dichloro-6,6'-dideoxy-sucrose, or 4,6,6'-trichloro-4,6,6'-trideoxy-galacto-sucrose in good yield, each crystalline or as a crystalline derivative, directly from the reaction without using chromatographic columns for isolation. Likewise, sulphonate esters of sucrose can be made directly, such as 1',6,6'-tri-O-mesitylenesulphonyl-sucrose.
We have studied the conversion of chlorosucroses into potentially useful products. Thus the di- and trichloro compounds are readily converted into 3,6:3', 6'-dianhydrosucrose and its 4-chloro derivative respectively. Furthermore, reductive dehalogenation of

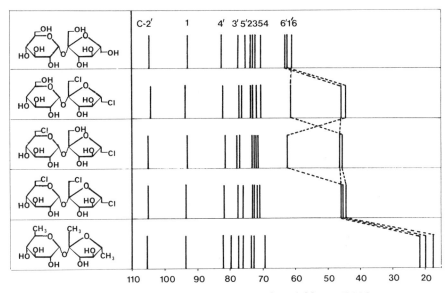

Figure 1. ^{13}C chemical shifts in ppm downfield from T.M.S.

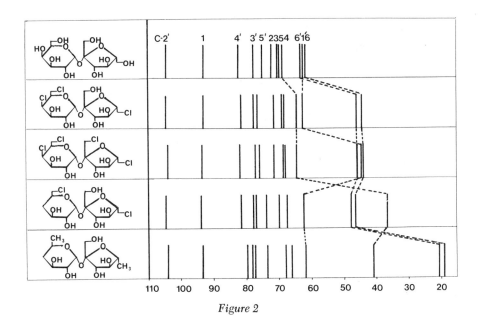

Figure 2

the chlorosucroses with hydrogen-Raney nickel in the
presence of base, afforded di-, tri- and tetradeoxy-
sucroses in high yield.
Studies on the application of ^{13}C n.m.r.to sucrose
and its derivatives have shown that it is a valuable
tool for characterisation.

Literature Cited

1.   Bolton, C.H., Hough, L. and Khan, R., Carbohyd.
     Res., (1972) 21, 133
2.   Hough, L. and Mufti, K.S. Carbohyd. Res., (1972)
     25, 497; (1973) 27, 47
3.   Richardson, A.C., Carbohyd. Res., (1969), 10, 395
4.   Khan, R., Carbohyd. Res., (1972), 25, 232
5.   Fairclough, P.H., Hough, L. and Richardson, A.C.,
     Carbohyd. Res. (1975), 40, 285
6.   Lindley, M.G.,Birch, G.G. and Khan, R.,J. Sci. Fd
     Agric., (1976), 27, 140
7.   Creasey, S.E. and Guthrie, R.D., J. Chem. Soc.
     Perkin, (1974) 1, 1373
8.   Almquist, R.G. and Reist, E.J.,J.Carbohyd. Nucleo-
     sides, (1974) 1, 461
9.   Hough, L.,Phadnis, S.P. and Tarelli, E.,Carbohyd.
     Res. (1975) 44, C12
10. Ballard, J.M. Hough, L. and Richardson, A.C.,
     Carbohyd. Res., (1974) 34, 184
11. Szarek, W.A.,Adv. In Carbohyd. Chem. and Biochem.
     (1973) 28, 225
12. Edwards, R.G, Hough, L.,Richardson, A.C. and
     Tarelli, E.,Tetrahedron Lett, (1973) 2369
13. Khan, R.,Jenner, M.R.,and Mufti, K.S. Carbohyd.
     Res., (1975) 39, 253
14. Parolis, H.,Carbohyd. Res., (1976) 48, 132
15. Ballard, J.M.,Hough, L.,Richardson, A.C. and
     Fairclough, P.H.,J. Chem. Soc., Perkin (1973)
     1, 1524
16. Hough, L.,Phadnis, S.P. and Tarelli, E.,Carbohyd.
     Res., (1975) 44, 37
17. Phadnis, S.P. and Hough, L.,unpublished results
18. Lawton, B.T.,Wood, D.J.,Szarek, W.A. and Jones, J.
     K.N.,Canad. J. Chem., (1969) 47, 2899
19. Hough, L.,Phadnis, S.P. and Tarelli, E.,Carbohyd.
     Res., (1975) 44, C12
20. Hough, L.,Khan, R. and Phadnis, S.P.,British
     Patent Applications No. 616/76 and 19570/76;
     Hough,L., and Phadnis, S.P., Nature,(1976)263,800.
21. Hough, L.,Phadnis, S.P. Tarelli, E. and Price, R.,
     Carbohyd. Res. (1976) 47, 151
22. Fukui, S. and Hayano, K.,Agr. Biol. Chem. (1976)
     33, 1013
23. Hough, L. and O'Brien, E. unpublished results.

Biographic Notes

Professor Leslie Hough, D.S., F.R.I.C., Prof.
Chem. Educated at Manchester Univ. and at Bristol
Univ. On staff at Bristol, leaving as Reader in 1967
to become Head of the Dept. of Chem. at Queen Eliza-
beth Coll. of the Univ. of London. Some 200 papers
in carbohydrate chemistry. Dept. of Chem., Queen
Elizabeth Coll., University of London, Atkins Building,
Campden Hill, London W.8. England.

3

High Resolution Nuclear Magnetic Resonance Spectroscopy

A Probe of the Structure and Solution Conformation of Sucrochemicals

L. D. HALL and K. F. WONG

Department of Chemistry, The University of British Columbia,
British Columbia, Canada V6T 1W5

W. SCHITTENHELM

Bruker-Spectrospin AG, Zurich Fallenden, Switzerland

During the past 6 or so years, there have been a number of developments in the area of high resolution, nuclear magnetic resonance (n.m.r.) spectroscopy which are pertinent directly to carbohydrate chemists who are interested in using n.m.r. to provide information concerning the structures and conformations of carbohydrate derivatives in solution. Two of these developments already have been used in studies of carbohydrate systems and in this presentation I intend to remind you (1-3) briefly of the significance of those techniques. First, the use of spectrometers which, by operating at high magnetic fields, provide an optimal dispersion of chemical shifts and thereby facilitate the use of the first three sets of n.m.r. parameters listed in Table I. Second, the use of spectrometers, which operate in the pulse Fourier transform mode and which, thereby, extend applications of the same three sets of parameters to solutions which magnetically are rather dilute, as in measurements of natural abundance (4) ^{13}C n.m.r. spectra. Since I assume that both of these areas are rather well known, I shall pass through this part of my presentation rather rapidly and then I shall go on to discuss an area which is rather newer and, I anticipate,less familiar to most carbohydrate chemists. This is the use of pulse Fourier transform methods to measure the fourth family of n.m.r. parameters listed in Table I, namely spin-lattice, or longitudinal relaxation times (T_1-values, sec); or their reciprocals, the spin-lattice relaxation rates (R_1-values, msec^{-1}). And I shall show (5-6) that proton R_1-values provide us with valuable new insight to the solution conformations of sugars in general,and disaccharides such as sucrose, in particular.

Table I.   High resolution n.m.r. parameters

1. Chemical Shifts (p.p.m.)
2. Coupling Constants (Hz)
3. Integrated areas
4. Spin-lattice

   Longitudinal } relaxation times (sec.)

5. Spin-spin

   Transverse } relaxation times (sec.)

(A)

The proton n.m.r. spectra of the tetrachloro-
tetramesyl galacto-sucrose derivative (A), which
are shown in Figure 1 serve to illustrate the advantage
of measuring ^{1}H n.m.r. spectra at as high a magnetic
field as possible.   The 100 MHz spectrum (Figure 1)
shows each of the ring protons as a separately resolved
resonance; however, the six methylene protons give a
poorly resolved multiplet.   In contrast, the 360 MHz
n.m.r. spectrum of the same solution now shows all of
these resonances clearly resolved and the assignments
of these and the various ring protons are given in
Figure 1.   With a spectrum which is dispersed thus, it
is a trivial matter to obtain all of the vicinal
^{1}H-^{1}H coupling constants and these values are below.

(A)

Following the pioneering observations of Lemieux (7) and the subsequent calculations of Karplus (8-9) it is a trivial matter to infer from these couplings that the pyranose ring of (A) has the D-galacto configuration and favours the $^4C_1$-chair configuration. Thus the couplings between H-1:H-2 and H-3:H-4 are characteristic of the gauche orientation, $J_{4,5}$ is typical of the D-galacto configuration and the large value of $J_{2,3}$ provides unequivocal proof that H-2 and H-3 have a trans-diaxial disposition. Turning now to the fructofuranose ring, we note that both the vicinal couplings are large and, although there are many reasons (10) why one cannot make an unequivocal conformational assignment, it is tempting to speculate that this ring favors the $^3T_4$-twist conformation in which C-3 is displaced above and C-4 equally below the reference plane defined by C-2:O-5:C-5. We shall return later to some additional evidence in support of this assertion.

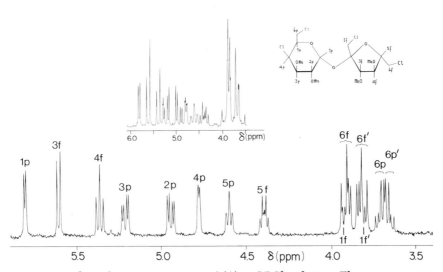

Figure 1. High resolution n.m.r. spectra of (A) in CDCl₃ solution. The upper spectrum was measured at 100 MHz using a Varian XL-100 instrument. The lower one was measured at 360 MHz using Bruker HX-360 instrument operating in the pulse Fourier transform mode.

Although these vicinal couplings clearly provide invaluable evidence concerning the conformations of the individual rings, it is noteworthy that nothing can be inferred as to the relative spatial disposition of the two saccharide units and we also shall return later to this point.

I shall now turn to the use of the pulse Fourier transform (F.t.) n.m.r. method. It will be recalled that, in a conventional continuous wave n.m.r. measurement, each nuclear resonance signal is excited in turn by a weak "observing" radiofrequency field which is scanned through the appropriate frequency region. Of necessity this is a slow process; typically 500 seconds is required to scan a 1000 Hz spectral region. In the pulse F.t. experiment, a rather strong "observing" radiofrequency transmitter is positioned to one side of the spectral region of interest and a short intense pulse of radiofrequency is applied in the form of a pulse. Without going into the details here (11), suffice it to say that the n.m.r. spectrum which is produced in this way represents a plot of magnetization intensity as a function of time. This often is referred to as a "free induction decay signal" and consists (Figure 2,part A)of a series of overlapping sine-waves, the frequencies of which represent the frequency separations from the transmitter frequency of each of the individual transitions in the n.m.r. spectrum. This rather unusual mode of display is then mathematically converted to the more familiar type of n.m.r. display (Figure 2,part B),which is a plot of magnetisation intensity versus frequency, using the theorem of Fourier. It is not inappropriate to note here that this transformation can now be performed rather rapidly by a small digital computer as a result of the Cooley-Tuckey algorithms (12).

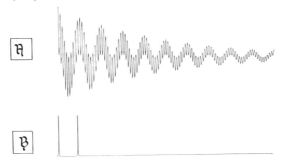

Figure 2. Examples of n.m.r. spectra displayed in both the time domain A, and the frequency domain B

In any event, the best known application of the
pulse F.t. experiment is that it enhances considerably
the speed with which an n.m.r. spectrum can be measured
and thereby enables spectra to be obtained from solutions which magnetically are rather dilute - and, for
the practicing organic chemist, the most important use
of this lies in the measurement of natural abundance
^{13}C n.m.r. spectra.   This is a formidable problem.
The ^{13}C nucleus is present naturally only in 1.1% abundance and, even if it were 100% abundant, its effective
sensitivity would only be ca. 1% that of an equivalent
number of protons.   Thus, there is an effective decrease in signal-to-noise of approximately 10^4 compared
with the proton.   Fortunately, the pulse F.t. method,
together with other important instrumental developments
such as noise-modulated decoupling, is equal to this
task.
    The ^{13}C spectra shown in Figure 3 were measured at
90.5 MHz; that in part A with simultaneous decoupling
of all of the protons and that in part B with the proton decoupler turned off.   Clearly visible in the
latter spectrum are the various ^1J proton couplings.

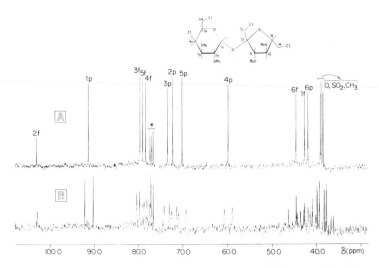

*Figure 3.   Natural abundance ^{13}C n.m.r. spectra of (A) in CDCl$_3$ solution
(0.1 molar in a 5-mm diameter n.m.r. tube). The spectrum in Part A represents the average of 5000 transients and was measured with simultaneous
coupling of all of the proton resonances. The spectrum of Part B was obtained with the decoupler gated off during the acquisition time, and again
5000 transients were used.*

The multiplicity of these resonances provides direct evidence for the number of directly bonded protons, and the magnitudes of the couplings reflect the orientation of the associated substitutent, at least at the anomeric centre (13). Now, although these spectra are rather impressive - indeed we believe that they are the first ^{13}C spectra of a sugar to be measured at 90.5 MHz - it is important to stress that, even had this spectrum been measured at 20 MHz, the dispersion between the individual transitions would have been adequate for our purposes. This is because ^{13}C chemical shifts are extremely sensitive to their chemical and stereochemical environment. Thus, for derivative (A) the entire proton spectrum covers a spectral region of only ca. 3.5 ppm whereas the ^{13}C spectrum spans ca. 70 ppm - an approximately twenty fold increase.

The true importance then, of measuring the ^{13}C spectra at high field is not the increased dispersion of the ^{13}C spectrum but, rather the increased dispersion of the proton spectrum. This is because it is necessary to have a method for assigning each of the ^{13}C resonances and it is obvious that, since each resonance is a singlet and there is no mutual [13]C-^{13}C coupling, this cannot be done by direct inspection. However, if the ^{1}H spectrum is sufficiently dispersed for each proton resonance to be assigned, then those ^{1}H assignments can be transferred to the ^{13}C resonances by selective proton decoupling. Briefly, a rather weak decoupling field is located in turn at the resonance frequency of each proton, and the ^{13}C spectrum measured in the normal fashion. Now, the observed spectra are a blend of those shown in Figure 3, part A and B - all of the resonances are detected as multiplets with the sole exception of that carbon which is directly coupled with the proton which is being selectively irradiated, which now shows as a sharp, spin-decoupled singlet. The assignments given above, Figure 3, part A, were all made in this way and are unequivocal. We believe that selective proton decoupling is the method of choice for assigning the ^{13}C spectra of sugars, simply because it is relatively simple to obtain adequate dispersion of their ^{1}H spectra.

I want to spend the remaining part of my presentation drawing attention to a second important application of the pulse Fourier transform technique - namely the measurement of the spin-lattice relaxation rates (R_1-values) of the proton and ^{13}C nuclei of carbohydrate derivatives in general, and of sucrochemicals in particular - which is the fourth of the five n.m.r. parameters listed in Table I.

In other studies (5), we have proven that the proton (14) and ^{13}C nuclei (15) of many carbohydrate derivatives relax exclusively by the dipole-dipole mechanism.   This relaxation mechanism has the general form
shown in equation One.This tells us that the efficiency
with which a "donor" nucleus (D) contributes to the
relaxation of a "receptor" nucleus (R) is proportional
to the square of the magnetogyric ratio (γ) of both D
and R - since the only nucleus in most carbohydrate derivatives which has a significant value of γ is the proton, it comes as no surprise to find that all the relaxation of proton (14) and ^{13}C nuclei (15) come from
the protons of the compound.   And, the fact that it
is the nearest neighbour protons which contribute
most to this relaxation, also is consistent with Equation One, according to which the relaxation efficiency
should fall off as the inverse sixth power of the distance between D and R.

Equation One.

$$R_1(D \rightarrow R) = \alpha \frac{\gamma_D^2 \cdot \gamma_R^2}{r_{D \rightarrow R}^6} \cdot \tau_c(D \rightarrow R)$$

In spite of their common genesis, proton R_1-values
provide quite different, albeit complimentary,structur-
al information than those of ^{13}C nuclei.   We have
developed (5) four different methods by which the magnitudes of individual, interproton relaxation contributions may be determined accurately.   It is obvious
that, if one determines the relative magnitudes of the
contributions which two donor protons D1 and D2 make
to the relaxation of a common receptor proton, R, this
gives a direct measure of the ratio of the interproton
separations, as depicted in Equation Two.   Clearly
this equation will be valid only if both of the inter-
nuclear vectors D1→R and D2→R have the same τ_c-value,
that is the same motional correlation time.....and this
is where the ^{13}C R_1-values enter the picture.   Because
of the inverse sixth power dependence of the dipole-dipole relaxation mechanism, the relaxation of ^{13}C nucleus will be dominated by the protons to which it is
directly bonded.   It follows then that, if every C-H
bond length is the same in a sugar derivative, the R_1-
values of each carbon bearing a single proton can be
identical only if each C-H vector has the same τ_c-value.
This can be so only if the molecule is tumbling iso-
tropically in solution.   Thus, measurement of ^{13}C R_1-

values provides a direct indication, not only of the
rate and nature of the tumbling motion of a sugar, but
also provides a splendid quality control experiment for
the relationship which is summarised by Equation Two.

Equation Two.

$$\frac{r^{D1 \to R}}{r^{D2 \to R}} = \sqrt[6]{\frac{R_1 (D2 \to R)}{R_1 (D1 \to R)}}$$

We shall return later to a brief discussion of one
of the methods whereby spin-lattice relaxation rates
may be determined routinely.   First, it is appropriate
to demonstrate that these parameters do have practical
relevance to the sucrochemical area.   The proton R_1-
values for (A) were determined at 360 MHz and the data
are given below (in msec^{-1}).   It will be recalled that
these composite R_1-values should be dominated by near-
est-proton interactions and, hence, those protons
which are closest together should relax the most rapidly, that is have the higher R_1-value.   In accord with
this expectation, the H-6p protons of the galactopyranose ring of (A) relax more rapidly than the other ring
protons of that moiety.   That H-2p relaxes the slowest
simply reflects the fact that it has only one nearest
neighbour interaction - with H-1p - plus a second, very
much, weaker interaction with H-3p.   Turning now to the
protons on the lower face of this ring we note that,
even though both H-3p and H-5p are also axially oriented, both relax more rapidly than H-2p.   In the case
of H-5p, this reflects the fact that it can pick up
relaxation from H-3p and the two H-6p protons.   But,
the reason for the enhanced relaxation of H-3p compared
with H-2p can be ascribed only to a trans-annular contribution received from H-5p.   From other studies (5,
16-18) we know that this trans-annular contribution is
rather substantial and it provides a very important
new probe for conformational studies of organic molecules.   In any event it is clear from the above, rather
qualitative, analysis that the proton R_1-values of
the galactopyranose moiety of (A) are completely consistent with the $^4C_1$ chair configuration previously assigned on the basis of the vicinal couplings.
    Turning now to the data for the fructofuranose
ring of (A) we note that once again the methylene protons relax more rapidly than the methine protons.   The
fact that the latter relax rather slowly certainly is

consistent with the previously assigned $^3T_4$ conformation and thus supports that previously tentative assignment.

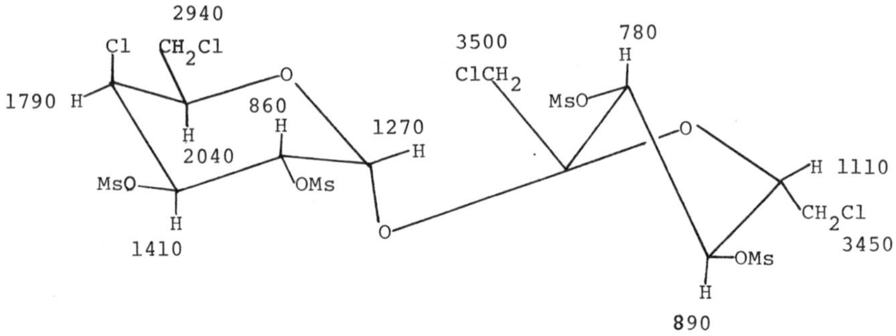

(A)

We now come to discuss the most important single
feature of the entire discussion so far.   Granted that
the relaxation of these protons accords with Equation
One, and that interproton relaxation contributions are
reciprocated between the donor and receptor nuclei, why
should the anomeric proton of the galactopyranose moie-
ty(H-1p) relax so much more rapidly than its neighbour,
H-2p?   If anything, one would have anticipated that,
since it is further away from the other protons of the
galactopyranose ring, it would relax more slowly.   The
obvious answer is that H-1p must receive additional relaxation contributions from elsewhere and the obvious
source is the proton substituents of the fructofuranose
ring.   If this be so, then this obviously opens up an
exciting new vista for conformational studies of sucrochemicals, and of any other derivatives which have two
or more continuous saccharide rings.

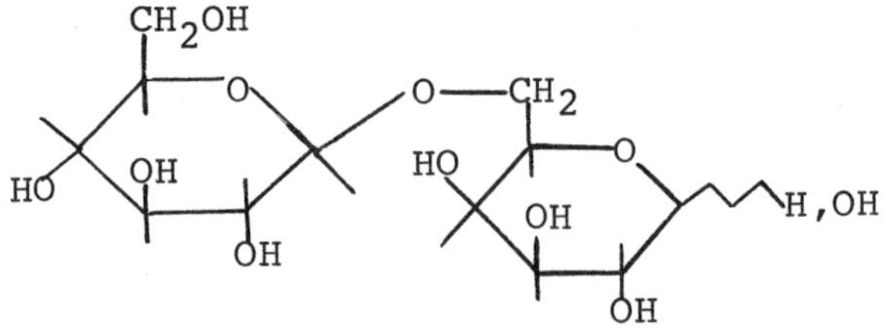

(B)

At present we have no experimental data for (A)
but data for the disaccharide derivative (B) (D-gluco-
pyranosyl-1 ⇄ 6-D-galactopyranose) proves (19) conclu-
sively that inter-ring, proton relaxation can be substantial (20).   Consider first the R_1-values for the
three anomeric protons of (B).   The differential between the relaxation rates of the anomeric protons at
the reducing centre, simply reflects the anticipated
(16-18) differential of axial and equatorial protons.
The axial proton of the β-anomer is closer to the axial
protons H-3 and H-5, than is the equatorial proton of
the α-anomer.   The large rate enhancement of the axial
proton at the non-reducing centre as compared with its
counterpart at the reducing centre must, according to
our speculation, arise from neighbouring protons on the
reducing ring.   Clearly the protons at C-6 of the
galactopyranose ring are the most probable donors.
Proof that this is so follows from measurement of the
proton R_1-values of the 6,6-dideutero derivative (C).
It will be recalled from Equation One that the relaxation efficiency of a donor nucleus is proportional to
the square of its magnetogyric ratio - which is why
protons are the dominant source of relaxation.   Since
the magnetogyric ratio of deuterium is only one sixth
that of the proton, it follows that replacement of a
proton by a deuterium atom effectively eliminates the
relaxation contribution from that site.   (Because
deuterium has a nuclear spin of unity whereas the
proton has a spin of one-half, the relaxation contribution from a deuterium nucleus is 6% that of a proton
located at precisely the same position.)   Thus comparison of the proton R_1-values of any derivative with
those of a specifically deuterated analog provides a
direct measure of a specific interproton relaxation
contribution.

(C)

Returning now to the disaccharides (B) and (C) we
note that the relaxation of the non-reducing anomeric
proton of (C) is substantially lower than that of its

counterpart in (B), which is as it should be. The
calculated value of the relaxation contribution between
H-1' and the two H-6 protons is 530 msec^{-1} at 30°,
which is 24% of the total relaxation rate.

Unfortunately time does not permit me to delineate
here any further details of our studies of these and
other glycoside and disaccharide derivatives. Suffice
it to say we have shown (19-21) that many 1→4- and
1→6-linked disaccharides have substantial inter-ring
relaxation contributions and we have used the specific
deuteration method to quantitatively identify the magnitudes of these interactions for a number of glycosides. These experiments provide the first direct
measure of the relative spatial disposition of the two
sugar rings of a disaccharide in solution. The
generality of this experimental approach and the importance of this type of structural information must be
quite obvious. In the case of sucrochemicals, it is
highly probable that the most important physiological
property of sucrose - namely its sweet taste - is in-
timately associated with the overall shape of the molecule and with the way in which it associates with the
taste-receptor protein. Thus, it would be of considerable interest to establish the overall conformation of sucrose in solution, to find out what this
shape has in common with the shapes of other molecules
which have a sweet taste, and how it differs from the
shapes of molecules which have either no taste or are
bitter. Furthermore, being mindful of the many important physiological properties of other oligosaccharides
- such as the oligosaccharide components of the blood
group substances, and the many amino-glycoside antibiotics - it must be obvious why my group at U.B.C. currently are, with considerable interest, pursuing many
experiments in these areas.

Anticipating that others will share our interest
in spin-lattice relaxation, and rather conscious of the
present shortage of simple accounts of spin lattice
relaxation, I shall end this presentation with a brief,
oversimplified account of this phenomenon and of one of
the several pulse F.t. methods which can be used to
measure spin lattice relaxation rates.

Spin lattice relaxation involves the interchange
of magnetic energy, magnetization, between the nuclei
of interest, the "spins" and their surroundings, the
"lattice". This energy exchange process involves
interactions between the spins and rapidly fluctuating
magnetic fields which are generated in the lattice by
the motion of various magnetic species (such as another
proton). Of the several possible mechanisms whereby

these fields can be generated, the most important one for the organic chemist is the "dipole-dipole" mechanism, which we have discussed earlier.

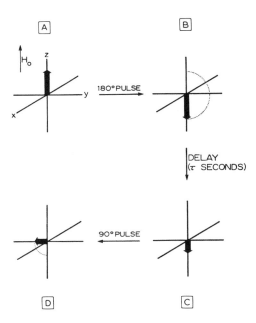

Figure 4. Diagrammatic representation of the rotating reference-frame model of a spin-lattice, relaxation-time measurement using a two-pulse sequence. (In A the magnetization of the nuclei is at thermal equilibrium with the lattice. In B this magnetization has been inverted through 180° by application of a 180°-pulse; the nuclei are no longer in thermal equilibrium, and the spin-lattice relaxation causes the magnetization to revert back along the z-axis towards its equilibrium position. After a known delay time—pulse-delay—the residual magnetization C is assayed by the application of a 90°-pulse which tips the magnetization up into the x-y-plane D.)

The spin-lattice relaxation rate of a particular set of equivalent nuclei is the first-order time constant of the energy exchange process for those nuclei and we now shall use the rotating reference frame model to illustrate one of several possible methods whereby these time constants may be measured. Consider our ensemble of spins. When they are at thermal equilibrium with the lattice, their net magnetisation can be represented as a vector directed along the +z-axis (see

Figure 4), which is the direction of the external applied magnetic field.   These nuclei can be effectively
"heated", and thus their thermal equilibrium destroyed,
by applying a pulse of energy at the correct frequency,
and having a total energy content sufficient to tip the
net magnetisation vector through 180°- this is referred to as a 180°-pulse.   At this point the process of
spin-lattice relaxation allows the nuclei to start to
"cool" down, back towards their equilibrium temperature
by passing their excess energy to the lattice.   In the
rotating reference frame, this is accompanied, first by
a progressive decrease in the magnetization intensity
along the -z-axis, followed subsequently by an increase
in the +z-direction.   And, if we follow this decay-re-
covery as a function of time, we have the required relaxation time constant.   In practice, a known period
of time (the delay time) is allowed to elapse following
the 180°-pulse and then the residual magnetization
which remains at the end of that period is measured by
inverting the magnetization up (or down) to the x, y-
plane by application of a 90°-pulse.   This induces,
for the first time in the experiment, a component of
magnetization which can be detected by the receiver of
the spectrometer (vide infra).   The sample is left
then for a sufficient time for the spins to reach thermal equilibrium with the lattice - generally a period
of "5 x T_1" is used, and then the above, two-pulse, in-
version-recovery sequence is used again but with a
different value of the delay time.   This gives a second datum point on the decay recovery curve.   Subsequently the cycle is repeated a further 10, or so,
times to define a complete relaxation curve for the
spins.
    The above two-pulse sequence has been used by phy-
sicists for many years, but with the advent of the
Fourier transform method it has become possible (22)
to use this method to measure simultaneously the relaxation rates of all the chemically shifted spins of a
complex molecule.   The 180°-pulse is applied simultaneously to all the nuclear spins and their residual magnetization is then measured using a non-selective 90°-
pulse.   Fourier transformation of the resultant free
induction decay curve gives the n.m.r. spectrum in the
required frequency domain.   By repeating the two-pulse
sequence with differing values for the pulse-delay, a
series of partially relaxed spectra are obtained
(see Figure 5).   Spectra measured immediately following the 180°-pulse have all transitions negative going,
whereas those measured after long delay times appear
as they would in a normal F.t. experiment, all the nu-

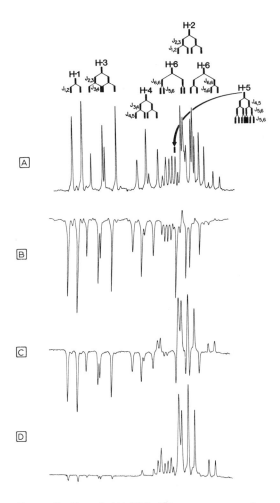

Figure 5. Partial, 100-MHz, ¹H n.m.r. spectra of methyl 2,3:4,5-di-o-isopropylidene-β-D-glucoseptanoside in benzene-d₆ solution (0.2M) using an acquisition time of 4.0 sec. (A and B were each obtained by using 9 transients, and C and D by using 16 transients. The spectrum shown in A is the normal spectrum. That in B was obtained by inverting the magnetization with a 180°-pulse and then sampling the residual magnetization after 1.0 sec by using a 90°-pulse. The spectra in C and D were obtained in the same way, but using pulse-delay times of 1.6 and 2.5 sec, respectively.)

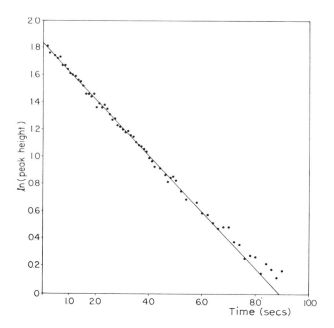

Figure 6. Typical decay plot of $(M_0\text{-}M_1)$ vs. t for the upfield transition of H-1α of a solution of D-glucose (10% w/v) in D_2O (99.96%) measured at 42°. (Note the "slowing down"—from purely exponential decay—at the longer delay times of ~1 T_1 period and greater.)

clei having reached full thermal equilibrium when the monitoring pulse was applied. The transition intensities for intermediate values of the delay time depend on the relationship between the delay time and the particular spin-lattice relaxation rate. Obviously nuclei with the shortest T_1-value, will relax more rapidly, hence, their magnetisation will pass the nulpoint and go positive considerably sooner than their more slowly relaxing neighbours. In any event, the relaxation rate is obtained by plotting a suitable function of the observed intensities versus delay time. If all is well, a single, smooth exponential curve will be obtained (see Figure 6).

With a modern pulse F.t. instrument it is a simple matter to perform the above experiment on a completely routine basis and no particular skills are needed.

Conclusions

It is appropriate to conclude this presentation with a brief summary. High resolution n.m.r. spectroscopy has established itself already as an extremely versatile method for studying the gross chemical structure of sucrochemicals, and future developments in both instrumental methods and theoretical concepts can only increase the scope and importance of the data that can be obtained. It has been possible to give here only a brief, somewhat arbitrary demonstration of a few of the methods which already are in common usage in n.m.r. spectroscopy, any of a wide variety of alternatives could equally well have been presented. It is clear that there is a pressing need for a close liaison between the sucrochemist and the n.m.r. spectroscopist and it is hoped that this presentation will encourage others to participate in such liaisons. From a personal standpoint, I now need some specifically deuterated sucrose derivatives and I would be most grateful if anyone could provide me with these. Equally good would be sucrochemicals which have either aldehyde, ketone or carboxylate substituents since we already know how to use those functional groups as precursors in deuteration synthesis. I note with much interest that Professor Hough already has a ketone derivative.

It remains to be seen whether there is a definite relationship between "shape" and "taste" for sucrochemicals and whether the spin-lattice relaxation method can provide useful information on this point.

Acknowledgements

It is a pleasure to thank Dr. John Hickson and the Internation Sugar Research Foundation for making it possible for me to participate in this important symposium, and Professor Leslie Hough for providing the sample of (A). The work performed at U.B.C. was supported by the National Research Council of Canada (grant A1905) to myself, and by a joint grant from NATO (No. 772) to myself and Professor L. Hough. K.F. Wong is a Commonwealth Scholar from Malaysia.

Abstract

High resolution n.m.r. spectroscopy long has been firmly established as a powerful tool for studying the structure and conformation of monosaccharide derivatives in solution. With the development of n.m.r. instruments which operate at high frequencies and in

the pulse Fourier transform mode, equivalent studies of
oligosaccharides should become equivalently routine,
and many totally new sources of structural data should
also evolve. Illustrative examples of some of this
potential in the sucrochemical area, derived from
studies in this laboratory, are discussed. Specifically: proton spectra measured at 360 MHz and ^{13}C
spectra at 91 MHz; assignments of ^{1}H spectra by a
pulse F.t. equivalent of the INDOR technique and of
^{13}C spectra via selective proton decoupling; and the
measurement of ^{1}H and ^{13}C spin-lattice relaxation
rates.

Literature Cited

1. Coxon, B., Advan. Carb. Chem., (1972), 27, 7.
2. Kotowycz, G., and Lemieux, R.U., Chem. Rev.,(1973)
    73, 669.
3. Hall, L.D., Advan. Carb. Chem., (1964), 19, 51.
4. Levy, G.C., and Nelson, G.L.,"Carbon-13 Nuclear
    Magnetic Resonance for Organic Chemists", Wiley
    Interscience, New York, N.Y., 1972.
5. Hall, L.D., Chem. Soc. Rev., (1975), 4, 401.
6. Hall, L.D. Chemistry in Canada, (1976), 28, 19.
7. Lemieux, R.U., Kulling, R.K., Bernstein, H.J.,
    and Schneider, W.G., J. Am. Chem. Soc., (1958),
    80, 6089.
8. Karplus, M., J. Chem. Phys., (1959), 30, 11.
9. Karplus, M., J.Am. Chem. Soc., (1963), 85, 2870.
10. Hall, L.D., Steiner, P.R., and Pedersen, C., Can.
    J. Chem., (1970), 48, 1155.
11. Farrar, T.C., and Becker, E.D., "Pulse and Fourier
    Transform N.M.R.", Academic Press, New York,
    N.Y., 1971.
12. Cooley, J.W., and Tukey, J., Math.Comput., (1965),
    19, 297.
13. Bock, K., and Pedersen, C., J.Chem. Soc. Perkin II,
    (1974), 293.
14. Hall, L.D., and Hill, H.D.W., J.Am. Chem. Soc.,
    (1976), 98, 1269.
15. Bock, K., and Hall, L.D., Carbohydr. Res., (1975),
    40, C3.
16. Preston, C.M., and Hall, L.D., Chem. Comm. (1972),
    1319.
17. Preston, C.M., and Hall, L.D., Carb. Res., (1974),
    37, 267.
18. Grant, C.W.M., Hall, L.D., and Preston, C.M.,
    J.Am. Chem. Soc., (1975), 95, 7742.
19. Berry, J., Hall, L.D., Welder, D.W.,and Wong,K.F.,
    unpublished results.

20. Hall, L.D., and Preston, C.M., <u>Carbohydr. Res.</u>, (1976), <u>49</u>, 3.
21. Berry, J., Hall, L.D., and Wong, K.F., unpublished results.
22. Vold, R.L., Waugh, J.S., Klein, M.P., and Phelps, D.E., <u>J. Chem. Phys.</u>, (1968), <u>51</u>, 3831.

Biographic Notes

    <u>Professor Laurance D. Hall</u>, Ph.D.; Prof. of Chem. Educated at Bristol Univ., post graduate fellow at Ottawa Univ. Joined the Univ. of British Columbia in 1964. Some 115 papers on carbohydrate chemistry, ^{13}C nmr and pnmr. Dept. of Chem., The Univ. of British Columbia, 2075 Wesbrook Place, Vancouver, B.C. V6T 1W5, Canada.

4

Some Fundamental Aspects of the Chemistry of Sucrose

RIAZ KHAN

Tate & Lyle Ltd., Group R & D, P.O. Box 68, Reading, RG6 2BX, Berks, England

Sucrose represents a regenerable chemical resource of undetermined commercial potential. Its value has been long recognized as a chemical raw material for food additives, surfactants, plastics and polymers, agricultural chemicals and pharmaceuticals. However, actual commercial success, so far, has been limited. During the past few years, it has become evident that no significant progress in this direction can be made without a profound knowledge of the fundamental chemistry of sucrose. Our objective, therefore, has been to study the basic chemistry of this molecule. This paper reviews some recent work on chemical modifications of sucrose with the hope that it will form a basis for commercial exploitation of such derivatives.

Structure

Sucrose is a non-reducing disaccharide which is systematically named β -$\underline{D}$-fructofuranosyl α -$\underline{D}$-glucopyranoside ($\underline{A}$),and the numbering of the carbon positions in the molecule is as shown in Figure 1. Sucrose contains eight hydroxyl groups,three of which are primary (C-1', C-6, and C-6') and the remaining five are secondary (C-2, C-3, C-3', C-4, and C-4'). The structure of sucrose has been established ·both by chemical (1-4) and enzymic (5-7) syntheses. Its structure also has been confirmed by physical methods such as, X-ray crystallography (8,9), neutron diffraction (10), and nuclear magnetic resonance spectroscopy (11-12).

Tritylation Reaction

The trityl group has been used widely for the blocking of primary hydroxyl groups in carbohydrate chemistry (13). Tritylation of sucrose usually is

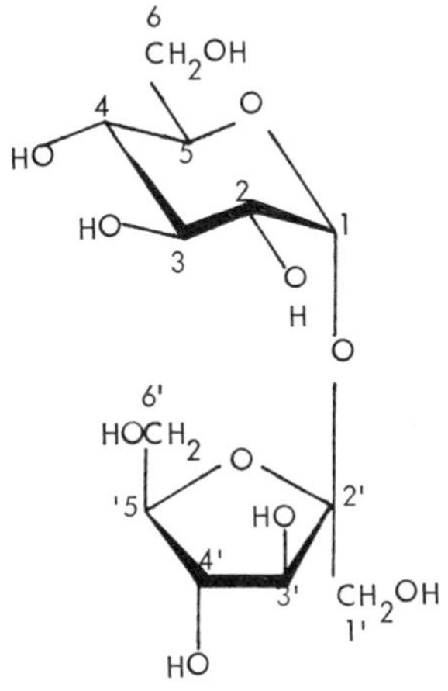

A   Sucrose

Figure 1

performed by reaction with an approximately stoichiometric amount of chlorotriphenylmethane in pyridine
at room temperature or above.   Tritylation at C-6 and
C-6' hydroxyl groups has been found to undergo completion in the order of 1 hour at 100°, whereas the more
hindered primary hydroxyl group at C-1' position reacts
more slowly.   Sucrose, on treatment with four molar
equivalents of chlorotriphenylmethane in pyridine for
48 hours at room temperature, after chromatography on
silica gel, afforded 6,6'-di-O-tritylsucrose (B) and
1',6,6'-tri-O-tritylsucrose (C) in 30 and 58% yield,
respectively (14).   The preferential reactivities of
the primary hydroxyl groups at C-6 and C-6' positions
in sucrose also has been observed in the transesterification reaction (15).   The tritylation reaction of
3,3',4',6'-tetra-O-acetylsucrose (D) has been described
(16).   Treatment of compound (D) with chlorotriphenylmethane and pyridine for 4 hours at 85°, gave 3,3',4',
6'-tetra-O-acetyl-6-O-tritylsucrose (E) and 3,3',4',6'-
tetra-O-acetyl-1',6-di-O-tritylsucrose (F) in yields of
67.7 and 14.8%, respectively.   When the reaction was
performed for 24 hours at 90°,the yield of compound
(F) increased to 85%.   (See Figure 2).
     Trityl ethers of sucrose are often crystalline
solids which are stable under basic and other nucleophilic conditions.   The trityl protecting group can be
removed under mild acidic conditions as, for example,
hydrogen bromide in glacial acetic acid or boiling
aqueous acetic acid.

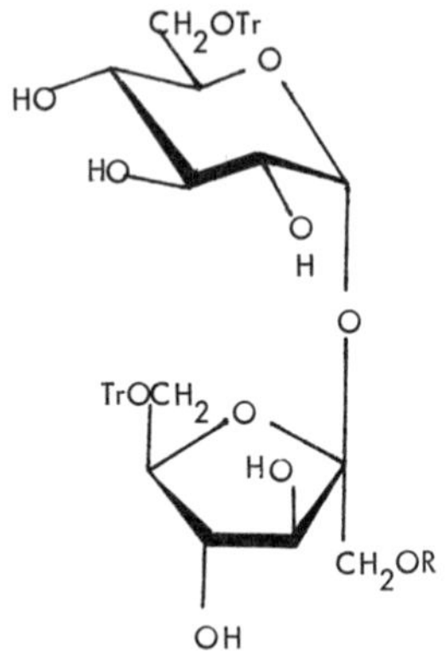

B R = H
C R = Tr

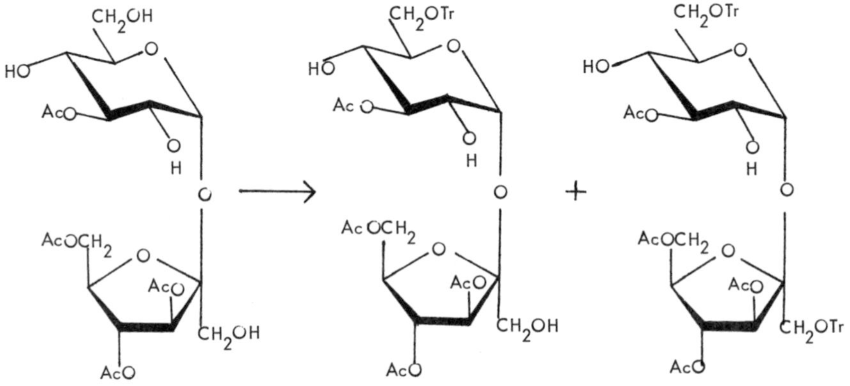

D   E   F

Figure 2

Methylation Reaction

Methylation of carbohydrates containing base-labile
substituents using a combination of diazomethane, dichloromethane, and boron trifluoride etherate is known
to proceed without concomitant migration of acyl groups
(17, 18).   Consequently, this method was chosen for
methylation of various, partially acylated derivatives
of sucrose (19).   Methylation of 1',2,3,3',4',6'-hexa-
O-acetylsucrose (G) with a freshly prepared solution of
diazomethane in dichloromethane and boron trifluoride
etherate for 0.5 hr at -5° after chromatographic separ-

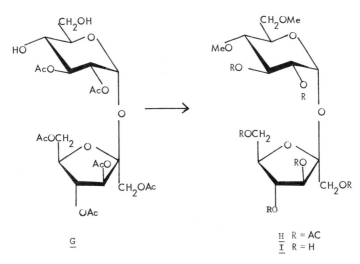

Figure 3

ation on silica gel,gave 1',2,3,3',4',6'-hexa-O-acetyl-
4,6-di-O-methylsucrose (H) in 95.5% yield (19). The
structure of (H) was supported by its 100 MHz proton
nuclear magnetic resonance (^{1}H n.m.r.) spectrum and by
mass spectrometry. The presence of a methyl group at
C-4 in (H) was indicated by the absence of an H-4 signal in the region of τ4.5-5.4 of the ^{1}H m.m.r. spectrum where it usually occurs in acetylated derivatives
of sucrose. the mass spectrum of(H)showed peaks at m/e
331 and 275 due to ketofuranosyl and hexopyranosyl cations, respectively. The free methyl ether (I) was
obtained in 67% yield, by treatment of (H) with a catalytic amount of sodium methoxide in methanol. (See
Figure 3).

Acylation Reaction

Acylation of sucrose generally is performed with
the appropriate acid anhydride or acyl halide in pyridine at or below room temperature. The most common
acyl derivatives of sucrose used are acetates and ben-

zoates.   Their value is well recognized in protecting
the hydroxyl groups of the sucrose molecule against
reactions which proceed under acidic and neutral conditions.   The cleavage of carboxylate ester protecting
groups can be effected under mildly basic conditions
such as catalytic sodium methoxide in methanol or
methanolic ammonia.

6,6'-Di-O-tritylsucrose (B), on treatment with
acetic anhydride and pyridine at room temperature, gave
the expected hexa-acetate(J) (14).   Detritylation of
2,3,3',4,4'-penta-O-acetyl-1',6,6'-tri-O-tritylsucrose
with hydrobromic acid in glacial acetic acid and chloroform at 0°, has been reported to give 2,3,3',4,4'-
penta-O-acetylsucrose in 74% yield (20).   The possibility of acyl migration must be taken into account in
the selective protection of sucrose.   When compound
(J) was treated with boiling aqueous acetic acid, detritylation accurred with concomitant migration of an acetyl group from O-4 → O-6, probably via the 4,6-ortho-
ester, to give 1',2,3,3',4',6-hexa-O-acetylsucrose (K)
(21).   In contrast, however, similar treatment of hexa-
O-benzoyl-6,6'-di-O-tritylsucrose and of 2,3,3',4,4'-
penta-O-benzoyl-1',6,6'-tri-O-tritylsucrose brought
about the expected losses of trityl groups but caused
little or no migration of ester groups (14, 22).
(See Figure 4).

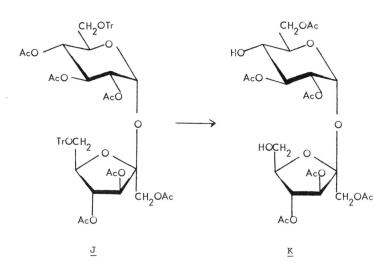

J   K

Figure 4

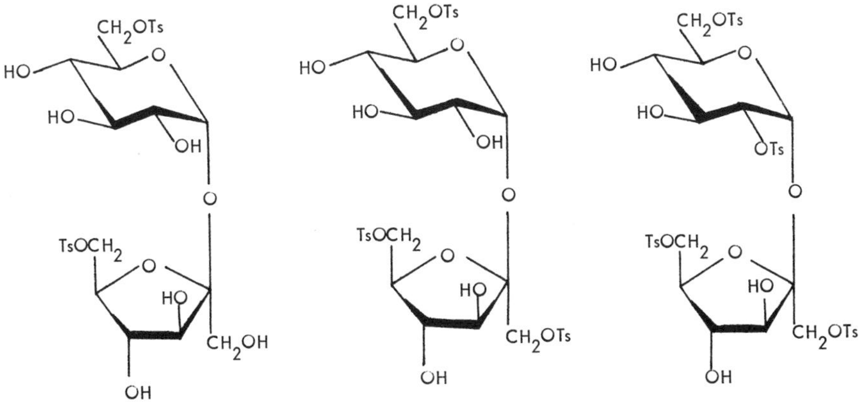

Figure 5

Sulfonylation Reaction

The selective p-toluenesulfonylation of sucrose has been investigated widely (22-27). Treatment of sucrose with about two molar equivalents of p-toluene-sulfonyl chloride in pyridine for 6 days at 0°, after chromatography on silica gel, gave crystalline 6,6'-di-O-p-tolylsulfonylsucrose (L) in 18% yield (26). A similar reaction of sucrose with three moles of p-toluenesulfonyl chloride and pyridine gave 1',6,6'-tri-O-p-tolylsulfonylsucrose (M) in 23% yield (22). Tetramolar P-toluenesulfonylation of sucrose has been reported to give, in addition to compound (M), 1',2,6,6'-tetra-O-p-tolylsulfonylsucrose (N) in 40% yield (27) (See Figure 5). The order of reactivity for sucrose is thus O-6≑ O-6'>O-1'>O-2. Methanesulfonylation reactions of sucrose and its derivatives also has been studied. Treatment of 1',2,3,3',4',6,6'-hepta-O-acetylsucrose at 0° gave crystalline 1',2,3,3',4',6,6'-hepta-O-acetyl-4-O-(methylsulfonyl)sucrose in 88% yield (28). The location of the isolated methanesulfonyloxy groups in sucrose can be identified by high resolution ^{1}H n.m.r. spectroscopy. In comparison with the ^{1}H n.m.r. data for octa-O-acetylsucrose, the signal due to H-4, in the above 4-sulfonate derivative, appeared at a slightly higher field, i.e., at τ5.2 (28). Similarly, in the ^{1}H n.m.r. spectrum of 1',3,3',4',6,6'-hexa-O-acetyl-2,4-di-O-(methylsulfonyl)sucrose, the signals due to H-2 and H-4 appeared at relatively higher fields, i.e., at τ5.35 and 5.2, respectively (16).

Acetalation Reaction

The value of cyclic acetals for the protection of
hydroxyl groups in carbohydrate chemistry is well known.
Several cyclic acetals of sucrose were successfully
synthesized in the last two years.   Sucrose, on treatment with benzylidene bromide and pyridine for 0.5 hour
at 95° , after acetylation and chromatographic separation, gave 1',2,3,3',4',6'-hexa-O-acetyl-4,6-O-benzyli-
denesucrose (O) in 35% yield (29).   The method involving the use of 2,2-dimethoxypropane, N,N-dimethylforma-
mide, and p-toluenesulfonic acid has been applied to
monosaccharides to afford strained and otherwise inaccessible cyclic acetals (30-33).   Use of this combination of reagents with sucrose and 6,6'-dichloro-6,6'-
dideoxysucrose (P) gave several acetals. Sucrose, on
treatment with 2,2-dimethoxypropane in N,N-dimethylfor-
mamide in the presence of a catalytic proportion of p-
toluenesulfonic acid,gave 4,6-O-isopropylidene- and 1',
2:4,6-di-O-isopropylidenesucrose in 55 and 15% yield,
respectively (16,34).   The latter product, apparently,
constitutes the first example in carbohydrate chemistry
of a compound with an eight-membered,cyclic acetal ring
(1',2-O-). The reaction of compound (P) with the above
combination of reagents, after acetylation and chromatography on silica gel, afforded 3,3',4,4'-tetra-O-ace-
tyl-6,6'-dichloro-6,6'-dideoxy-1',2-O-isopropylidene-
sucrose (Q) and 3',4'-di-O-acetyl-6,6'-dichloro-6,6'-
dideoxy-1',2:3,4-di-O-isopropylidenesucrose(R)in yields
of 37 and 39%, respectively (35) (See Figure 6).   The
reaction of 1,6,6'-tri-O-tritylsucrose with 2,2-dimeth-
oxypropane--N,N-dimethylformamide--p-toluenesulfonic
acid at room temperature, after acetylation and chromatography on silica gel, gave 2,3-O-isopropylidene-1',
6,6'-tri-O-tritylsucrose tri-acetate and 3,4-O-isopro-
pylidene-1',6,6'-tritylsucrose tri-acetate in yields of
24% and 20%, respectively (35).
     These results indicate that the order of preference
for the formation of acetals in sucrose is 4,6>2,1'>2,3
>3,4.

Bimolecular Nucleophilic Displacement Reaction

Bimolecular nucleophilic displacement reactions of
sucrose sulfonates and deoxyhalides have been studied.
Treatment of octa-O-(methylsulfonyl)sucrose (S) with
sodium iodide in butanone under reflux gave 6,6'-di-
deoxy-6,6'-dideoxy-6,6'-diiodo-1',2,3,3',4,4'-hexa-O-
(methylsulfonyl)sucrose (T) in 75% yield (26).   The

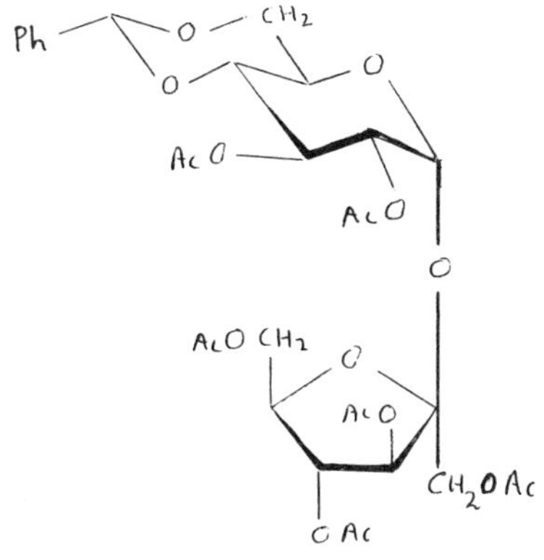

Figure 6

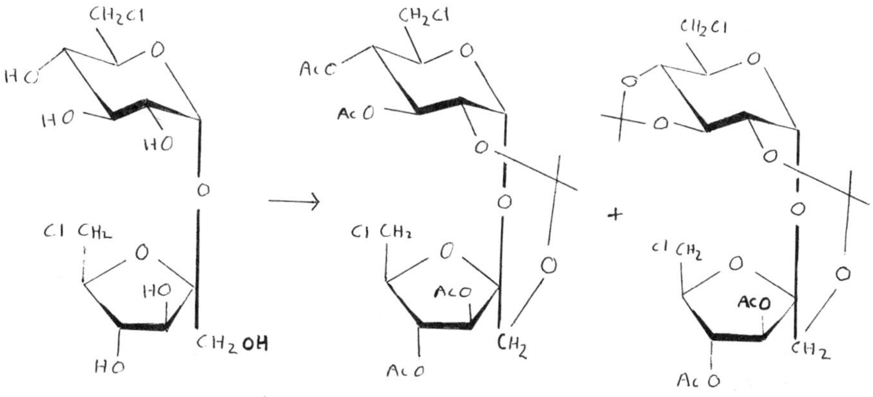

inactivity of the sulfonyloxy group at C-1' in (S) probably is due to unfavorable steric and polar factors.
A difference in the reactivities of the sulfonyloxy
groups at C-6 and C-6' was noted during the reaction of
6,6'-di-O-p-tolysulfonyl sucrose,   benzoyl chloride,
and pyridine at room temperature (26).   Subsequent
investigation of the reaction of 1',2,3,3',4,4'-hexa-O-
acetyl-6,6'-di-O-p-tolysulfonyl sucrose with sodium
chloride in hexamethylphosphoric triamide, gave 1',2,
3,3',4,4'-hexa-O-acetyl-6, 6'-dichloro-6,6'-dideoxysu-
crose and 1'2,3,3 4,4'-hexa-O-acetyl-6-chloro-6-deoxy-
6'-O-p-tolylsulfonylsucrose in yields of 51 and 34%,
respectively (36).   This suggested that the sulfonyl-

48   SUCROCHEMISTRY

oxy group at C-6 is more reactive than the sulfonyloxy
group C-6'. A similar difference in the reactivities
of the sulfonyloxy groups at C-1' and C-4 in (S) also
has been established (37). Compound (S), on reacting
with sodium azide in hexamethylphosphoric triamide for
16 hrs. at 85°,gave 6-azido-6-deoxy-1,3,4-tri-O-(methylsulfonyl)-β-D-fructofuranosyl 4,6-diazido-4,6-dideoxy-
2,3-di-O-(methylsulfonyl)-α-D-galactopyranoside (U) in
80% yield. When the above reaction was performed for
48 hours at 90° it gave, in addition to the triazide(U)
(60%),1,6-diazido-1,6-dideoxy-3,4-di-O-(methylsulfonyl)-
β-D-fructofuranosyl 4,6-diazido-4,6-dideoxy-2,3-di-O-
(methylsulfonyl)-α-D-galactopyranoside (V)in 10% yield.
The secondary sulfonyloxy group at C-4 thus is more
reactive than that at the primary C-1' position in (S)
(See Figure 7).

Internal Displacement Reaction

    Sucrose sulfonates and deoxyhalides, under basic
conditions, have been shown to undergo internal displacement reactions to give anhydro derivatives (22,24,
26,38-42). The conversion of 6,6'-di-O-p-tolylsul-
fonylsucrose (L) into 3,6:3',6'-dianhydrosucrose (W)
has been achieved using sodium methoxide in methanol
(26). The dianhydride (W) also has been synthesized
from 6,6'-dichloro-6,6'-dideoxysucrose (P) in 80% yield
(43), by means of sodium methoxide in methanol. A
similar treatment of 1',2,3,3',4,4',6-hepta-O-acetyl-
6-chloro-6-deoxysucrose (X) with M sodium methoxide in
methanol under reflux, after conventional acetylation
with acetic anhydride and pyridine, gave crystalline
1',2,3',4,4',6-hexa-O-acetyl-3,6-anhydrosucrose (Y) in
83% yield (42).   (See Figure 8).

Elimination Reaction

    Dehydrohalogenation of 1',2,3,3',4,4'-hexa-O-ben-
zoyl-6,6'-dideoxy-6,6'-diiodosucrose, 1',2,3,3',4,4'-
hexa-O-benzoyl-6,6'-dibromo-6,6'-dideoxysucrose,
and 6,6'-dideoxy-6,6'-diiodo-1',2,3,3',4,4'-
hexa-O-(methylsulfonyl)sucrose, by means of silver
fluoride and pyridine gives the corresponding 5,5'-
diene derivative (26,36). 1',2,3,3',4,4',6-Hepta-O-
acetyl-6-deoxy-6-iodosucrose (Z) on treatment with
anhydrous silver fluoride in pyridine for 24 hours at
room temperature,after purifying from a column of silica gel, gave 1,3,4,6-tetra-O-acetyl-β-D-fructofuranosyl
2,3,4-tri-O-acetyl-6-deoxy-α-D-xylo-hex-5-enopyranoside
(AA) in 60% yield (44).   The high resolution ¹H n.m.r.
spectrum of (AA) showed the well-known (45) allylic

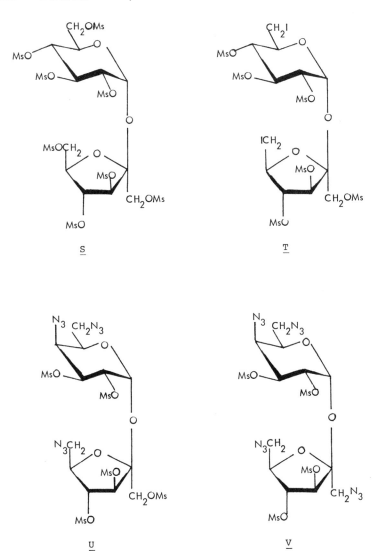

Figure 7

coupling between the protons at C-4 and C-6 (**44**).
(See Figure 9)

Reduction Reaction

Deoxy derivatives of sucrose have been prepared by
catalytic reduction of the corresponding deoxyhalides
or exocyclic vinyl ethers (**20,26,37,44**). Reduction of
compound (**Z**) in methanol with Raney nickel and hydrazine

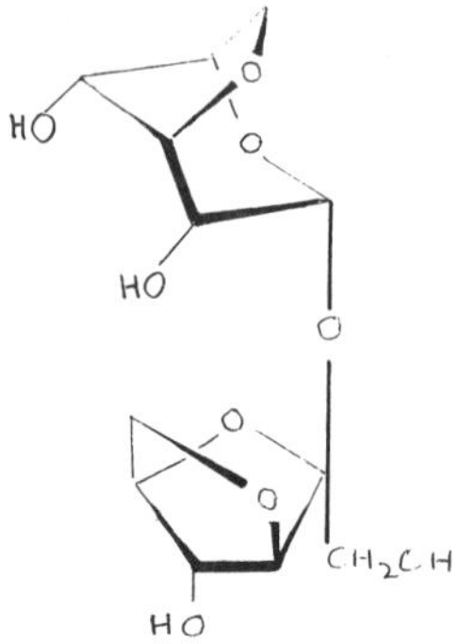

Figure 8

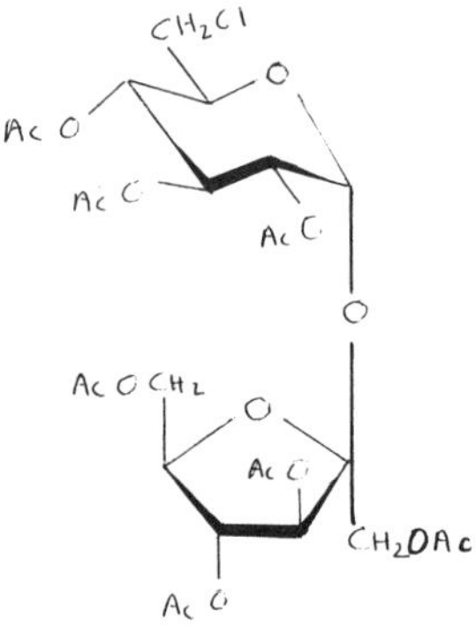

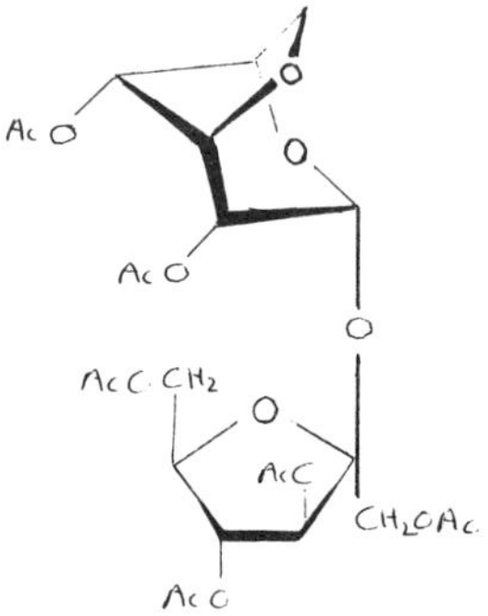

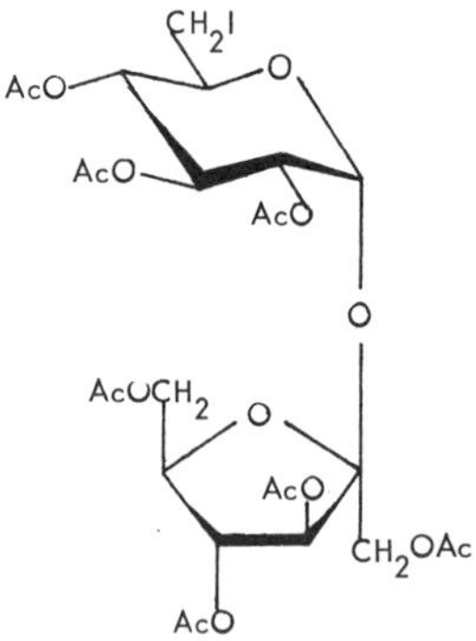

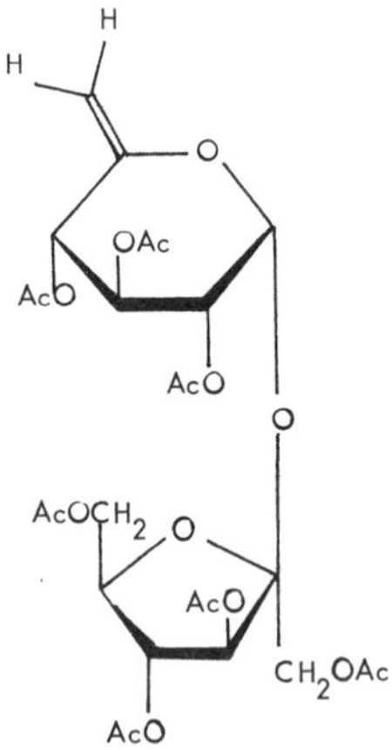

Figure 9

hydrate gave the expected 1',2,3,3',4,4',6'-hepta-O-
acetyl-6-deoxysucrose (BB) in 75% yield (44). Reduction of the exocyclic vinyl ether, β-D-fructofuranosyl
α-D-xylo-hex-5-enopyranoside by means of palladium-on-
charcoal in methanol, after acetylation and chromatography on silica gel, gave (BB) and 1,3,4,6-tetra-O-
acetyl-β-D-fructofuranosyl 2,3,4-tri-O-acetyl-6-deoxy-
β-L-idopyranoside (CC) in 10 & 46% yields respectively.
When similar reduction was performed with (AA) it gave
the L-ido isomer (CC) as the only isolable product in
46% yield. Little or no formation of the D-gluco isomer
(BB) was observed. The structure of (CC) was supported
by its 100 MHz H n.m.r. spectrum. The derived first-
order coupling constants ($J_{1,2}$2.0,$J_{2,3}$3.5,$J_{3,4}$3.5 and
$J_{4,5}$2.5Hz), were in agreement with theβ-L-ido configu-
ation and $^1C_4$ conformation for the hexopyranosyl residue
in (CC) (44). (See Figure 10).

Methanesulfonyl chloride -- N,N-Dimethylformamide Complex Reaction

     The methanesulfonyl chloride and N,N-dimethylfor-
mamide complex [Me$_2$N=CHO Ms] $^+$Cl$^-$ has been used for the
selective replacement by chlorine (46) of primary hydroxyl groups of hexopyranosides. Subsequent investigation of this reaction with methyl glucopyranosides,
methyl β-maltoside,and sucrose, revealed that substitution also occurs of secondary positions (42,47,48).The
substitution of the hydroxyl group at a chiral centre
has been shown to proceed with inversion of configuration (47). Treatment of 1',2,3,3',4',6'-hexa-O-acetyl-
sucrose (DD) with methanesulfonyl chloride and N,N-di-
methylformamide, initially for 2 hours at 0° and then
for 24 hr at 98°, gave two products which were separated on silica gel and characterized as 1,3,4,6-tetra-
O-acetyl-β-D-fructofuranosyl 2,3-di-O-acetyl-4,6-dich-
loro-4,6-dideoxy-α-D-galactopyranoside (EE) and 1',2,3,
3',4',6'-hexa-O-acetyl-6-chloro-6-deoxy-4-O-formylsu-
crose (FF) (42). The formation of formic esters has
been recognized (46) during the reaction of hexopyranosides w/methanesulfonyl chloride--N,N-dimethylformide
complex via hydrolysis of [Me$_2$N=CHOR]$^+$Cl$^-$. Our attempts
to isolate such intermediates recently have been successful (42). The formyl group in (FF) was selectively
cleaved by treatment with IRA-94S (HO$^-$) resin in methanol to afford 1',2,3,3',4',6'-hexa-O-acetyl-6-chloro-6-
deoxysucrose (GG). (See Figure 11).

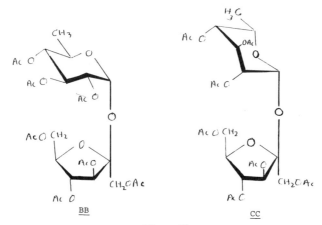

Figure 10

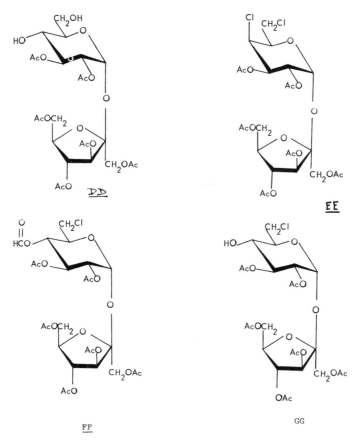

Figure 11

Chlorosulfonylation Reaction

The reaction of sulfuryl chloride with sugars has
been investigated to give products in which the primary
hydroxyl groups were replaced by chlorine and secondary hydroxyl groups either were esterified by cyclic
sulfate or substituted by chlorine with inversion of
configuration (49-57). The reaction of methyl β -D-glu-
copyranoside with sulfuryl chloride and pyridine has
been reported to proceed via methyl β-D-glucopyranoside
tetrakis(chlorosulfate) to give methyl 4,6-dichloro-4,6-
dideoxy-β -D-galactopyranoside 2,3-sulfate (54). The
assumed S_N2 character of the displacement reaction has
been questioned recently by Khan (55). Treatment of
2,3,3',4,4'-penta-O-benzoylsucrose (HH) with sulfuryl
chloride and pyridine in chloroform at -75° gave the
corresponding 1',6,6'-tris(chlorosulfate) (II) in 78%
yield. Compound (II) on treatment with pyridinium
chloride in chloroform for 4 hours at 50° afforded 2,3,
3',4,4'-penta-O-benzoyl-6,6'-dichloro-6,6'-dideoxysu-
crose 1'-chlorosulfate (JJ) in 76% yield. The value
of chlorosulfate residue as a leaving group has been
emphasized (54). However, when (II) and 1',2,3,3',4,4'-
hexa-O-benzoylsucrose 6,6'-bis(chlorosulfate)was treated with sodium azide in butanone,the only isolable products were (JJ)(83%) and 1',2,3,3',4,4'-hexa-O-benzoyl-
6,6'-dichloro-6,6'-dideoxysucrose (69%), respectively.
The above reaction suggested an effective competition
by the chloride ion which could have arisen only from
the chlorosulfate groups of (II) and 1',2,3,3',4,4'-
hexa-O-bensoylsucrose 6,6'-bis(chlorosulfate).
Hence, the displacement of the chlorosulfate groups in
(II) and 1',2,3,3',4,4'-hexa-O-benzoylsucrose 6,6'-bis
(chlorosulfate) by chloride ion, probably involved an
intramolecular process, similar to the S_N1 reaction of
alkyl chlorosulfites or alkyl chloroformates. (See
Figure 12).

Selective De-esterification of Sucrose Derivatives

Selective de-esterification of sucrose octa-ace-
tate, using a column of alumina (Laporte Type H), has
been reported to give 2,3,4,6,1',3',4'-hepta-O-acetyl-
sucrose, 2,3,6,1',3',4',6'-hepta-O-acetylsucrose, and
2,3,4,6,1',3',6'-hepta-O-acetylsucrose in yields of 9,
2.7 and 6% respectively (58). We have investigated
the use of ammonia in methanol to achieve selective de-
esterification of 2,1':4,6-di-O-isopropylidenesucrose
tetra-acetate (KK). Treatment of the diacetal (KK)
with ammonia in methanol at 0°, after chromatography on

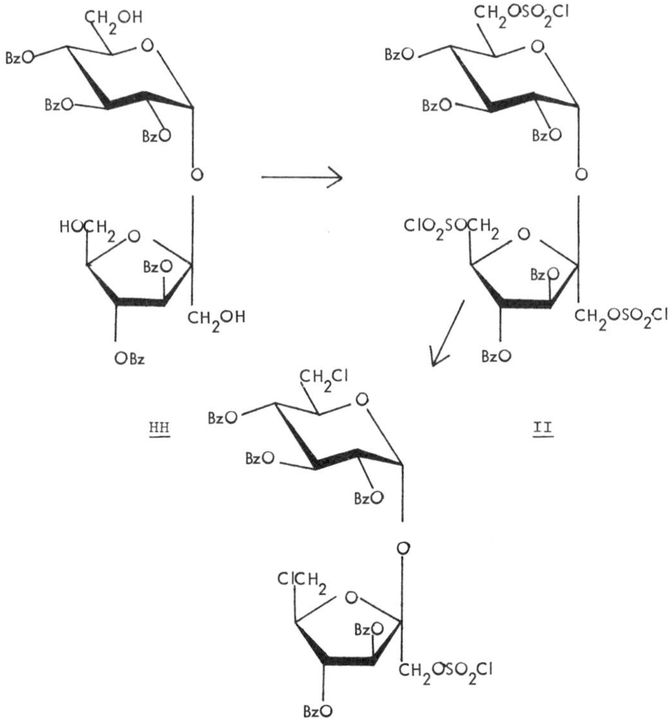

Figure 12

a column of silica gel, gave 3,6'-di-O-acetyl-2,1':4,6-di-O-isopropylidenesucrose (LL) and 3-O-acetyl-2,1':4,6-di-O-isopropylidenesucrose (MM) in yields of 71% and 22%, respectively (57). These results suggested that the lability of the acetyl groups in the furanose ring of (KK) were in the order of O-3=O-4>O-6. Lack of reactivity of the acetyl group at O-3 in (KK) probably is due to the steric effect of the neighbouring acetal (2,1':4,6-di-O-) groups. (See Figure 13).

Sucrose Epoxides

The value of sugar epoxides as synthetic intermediates is well-known in carbohydrate chemistry. Synthesis of sucrose epoxides hitherto has not been repor-

ted. However, an epoxide derivative of sucrose, 4,6-
dichloro-4,6-dideoxy-2,3-O-sulfo-α-D-galactopyranosyl
3,4-anhydro-1,6-dichloro-1,6-dideoxy-β-D-ribo-hexulo-
furanoside formed during the reaction of sucrose and
sulfuryl chloride at room temperature (57). The structure of the above epoxide has not yet been confirmed.

Intramolecular S_N2 reactions of 3',4'-di-O-p-tolu-
enesulfonylsucrose hexa-acetate (NN),3'-O-p-toluenesul-
fonylsucrose hepta-acetate(RR) and 4'-O-p-toluenesul-
phonylsucrose hepta-acetate(QQ) have been investigated
(59). Treatment of (NN) with M sodium methoxide in
methanol under reflex for 1-2 min, after conventional
acetylation and chromatography on silica gel, gave 2,3,
4,6-tetra-O-acetyl-α-D-glucopyranosyl 1,6-di-O-acetyl-
3,4-anhydro-β-D-lyxo-hexulofuranoside (OO) and 2,3,4,
6-tetra-O-acetyl-α-D-glucopyranosyl 1,6-di-O-acetyl-3,
4-anhydro-β-D-ribo-hexulofuranoside (PP) in yields of
76% and 12%, respectively. The structures of (OO) and
(PP) have been confirmed by unambiguous syntheses.
Reaction of the 3'-tosylate(RR) with M sodium methoxide
in methanol for 1-2 min, after acetylation and chromatographic fractionation, gave the ribo-epoxide (PP) in

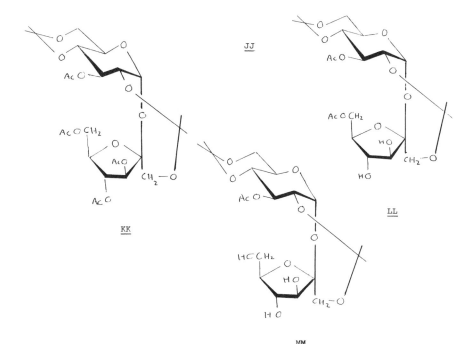

Figure 13

rise to a ring opening reaction exclusively at C-4' in each case (59). Consequently, reaction of the 3',4'-ribo-epoxide (PP) with sodium azide followed by acetylation gave α-D-glucopyranosyl 4-azido-4-deoxy-β-D-xylo hexulofuranoside hepta-acetate (SS) in 82% yield. Similarly,treatment of the 3',4'-lyxo-epoxide (OO) with sodium azide followed by acetylation gave, 4'-azido-4'-deoxysucrose hepta-acetate(TT) in 63% yield, as the only product. The direction of cleavage of these epoxides, presumably, were governed by a combination of 96% yield. The short time required for the formation of the epoxide indicated that the reacting groups(C-4',-OTs and C-3',-OAc)in (RR) had coplanar configurations. Synthesis of the lyxo-epoxide (OO)was achieved in 82% yield by treatment of the 4'-tosylate (QQ) with M sodium methoxide in methanol under reflux for 1-2 min,followed by acetylation with acetic anhydride and pyridine. (See Figure 14).

Ring Opening Reactions of Sucrose Epoxides

Treatment of the ribo-(PP) and lyxo-(OO) epoxides with sodium azide in aqueous ethanol was found to give steric and polar factors. In particular, the lack of reactivity at C-3' in each of these cases is indicative of the polar interactions which would occur between the permanent dipoles of the two C-2'-O bonds and the dipole which would be set up in the transition state. This can be considered to be analogous to the low reactivity of a 2-sulphonyloxy group, in S_N2 reactions, in hexopyranosides. (Figure 15).

Acknowledgment

I thank Professor A.J. Vlitos, Chief Executive of the Tate & Lyle Research Centre, for his interest and support, and K.S. Mufti and M.R. Jenner who have participated during the past few years in the program of work reviewed in this paper.My thanks are also extended to Professor L. Hough, Drs. K.J. Parker and H.F. Jones for helpful discussions.

Abstract

During the past six years, the objectives of our research have been to develop new applications of sucrose, by studying the fundamental chemistry of sucrose and its derivatives. The value of cyclic acetal groups is well recognised for the protection of hydroxyl functions in carbohydrate chemistry. The first synthesis of

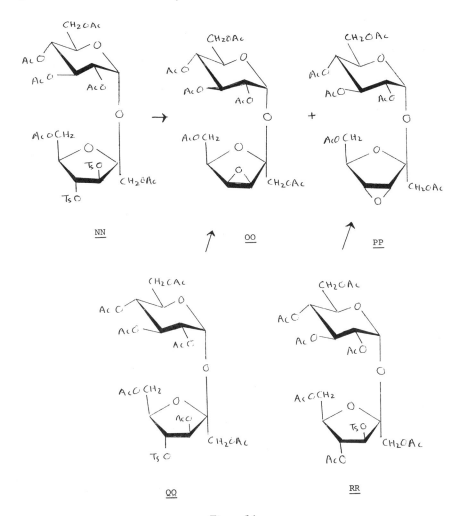

Figure 14

a cyclic acetal derivative of sucrose, 4,6-O-benzyli-
dene sucrose, was achieved in this laboratory in 1974.
Since then progress in this field has been rapid and
various cyclic acetals of sucrose and its derivatives
have been prepared.

Sucrose on treatment with 2,2-dimethoxypropane-N,
N-dimethylformamide-toluene-p-sulfonic acid (reagent A)
gave 4,6-O-isopropylidene (55%) and 1',2:4,6-di-0-iso-
propylidene derivative (15%).   When 6,6'-dichloro-6,
6'-dideoxysucrose was treated with reagent A, the corresponding mono-(1',2-) and di-(1',2:3,4-) acetals were

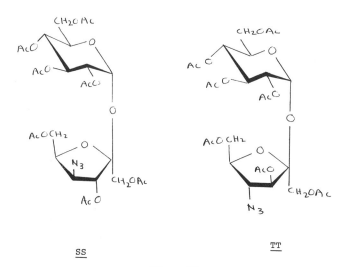

$\underline{\underline{SS}}$ $\underline{\underline{TT}}$

Figure 15

isolated in yields of 40% and 37%, respectively.   On
the basis of these results, it has been concluded that
the order of preference for the formation of acetal lin-
kage in the sucrose skeleton is 4,6>1',2>2,3>3,4. These
cyclic acetals of sucrose and their derivatives allow
selective reactions with previously inaccessible hydroxyl groups, in particular C-2, C-3, C-3', and C-4'. Some
of these selective reactions also will be discussed.

Literature Cited
1. Lemieux, R.U., Huber, G., J.Amer.Chem.Soc., (1953)
     75, 4118.
2. Tushida, H., Komoto, M., Agr. Biol. Chem. (Tokyo),
     (1963), 29, 239--242.
3. Ness, R.K., Fletcher, Jr., H.G., Carbohyd. Res.,
     (1971), 17, 465--470.
4. Reid, B.F., Intern. Symp. Carbohyd. Chem. VII in,
     Bratislava, (1974).
5. Hassid, W.Z., Dondoroff, M., Barker, H.A., J. Amer.
     Chem. Soc., (1944), 66, 1416--1419.
6. Hassid, W.Z., Dondoroff, M., Advan. Enzymol., (1950)
     10, 123--143.
7. Kauss, H., Z. Naturforsch., (1962), 176, 698--699.
8. Beevers, C.A., Cochran, W., Proc. Roy. Soc. (London)
     Ser. A., (1947), 190, 257--272.
9. Beevers, C.A., McDonald, T.R.R., Robertson, J.H.,
     Stern, F., Act. Crystallogr., (1952), 5, 689--
     690.

10.Brown, G.M., Levy, H.A., Science, (1963), 141, 921--
    923.
11.Binkley, W.W., Horton, D., Bhacca, N.S., Carbohyd.
    Res., (1969), 10, 245--258.
12.De Bruyn, A., Anteunis, M., Verhegge, G., Carbohyd.
    Res. (1975), 42, 157--161.
13.Helferich, B., Advan. Carbohyd. Chem., (1948), 3,
    79--106.
14.Hough, L., Mufti, K.S., Khan, R., Carbohyd. Res.
    (1972), 21, 144--147.
15.Reinfeld, E., Klaudianos, S., Zucker, (1968), 21,
    330--338.
16.Khan, R., Mufti, K.S., Carbohyd. Res., (1975), 43,
    247--253.
17.Gros, E.G., Flematti, S.M., Chem. Ind. (1966),
    (London) 1556-1557.
18.Deferrari, J.O., Gros, E.G., Mastronardi, I.O.,
    Carbohyd. Res., (1967), 4, 432--434.
19.Lindley, M.G., Birch, G.G., Khan, R., Carbohyd.Res.,
    43, 360--365 (1975).
20.Suami, T., Kato, N., Kawamura, M., Nishimura, T.,
    Carbohyd. Res., (1971), 19, 407--411.
21.Khan, R., Carbohyd. Res., (1972), 25, 232--236.
22.Khan, R., Carbohyd. Res., (1972), 22, 441--445.
23.Bragg, P.B., Jones, J.K.N., Can. J. Chem., (1959),
    37, 575--578.
24.Lemieux, R.U., Barrette, J.P., Can. J. Chem., (1959)
    37, 1964--1969.
25.Neuman, R., Ibara, J.A., J. prakt. Chem., (1972),
    314, 365--366.
26.Bolton, C.H., Hough, L., Khan, R., Carbohyd. Res.,
    (1972), 21, 133--143.
27.Ballard, J.M., Hough, L., Richardson, A.C.,
    (Unpublished Data).
28.Khan, R., Unpublished data.
29.Khan, R., Carbohyd. Res., (1974), 32, 375--379.
30.Evans, M.E., Parrish, F.W., Tetrahedron Lett., (1966)
    3805--3807.
31.Evans, M.E., Parrish, F.W., Long, Jr., L., Carbohyd.
    Res., (1967), 3, 453--462.
32.Hasegawa, A., Fletcher, Jr., G.H., Carbohyd. Res.,
    (1973), 29, 209--222.
33.Hasegawa, A. and Nakajima, M., Carbohyd. Res.,
    (1973), 29, 239--245.
34.Khan R., Mufti, K.S., Brit. Pat. 1,437,048.
35.Khan R., Jenner, M.R. and Jones, H.F., Carbohydrate
    Research, 49, 259--265 (1976); Khan R., and
    Jenner, M.R., (Unpublished data).
36.Hough, L., Mufti, K.S., Carbohyd. Res., (1972),
    25, 497--503.

37.Hough,L., Mufti, K.S., Carbohyd. Res., (1973),
   27, 47--54.
38.Lemieux, R.U., Barrette, J.P., J. Amer. Chem. Soc.,
   (1958), 80, 2243--2246.
38a.Lemieux, R.U., Barrette, J.P., Can. J. Chem.,
   (1960), 38, 656--662.
39.Buchanan, J.G., Cummerson, D.A., Turner, D.H.,
   Carbohyd. Res., .(1972), 21, 283--292.
40.Buchanan, J.G., Cummerson, D.A., Carbohyd. Res.
   (1972), 21, 293--296.
41.Isaacs, N.W., Kennard, C.H.L., O'Donnell, A.W.,
   Richards, G.N., Chem. Commun., (1970), 360.
42.Khan, R., Jenner, M.R., Mufti, K.S., Carbohyd.
   Res., (1975), 39, 253--262.
43.Khan, R., (unpublished data).
44.Khan, R., Jenner, M.R., Carb. Res.,48,306-311(1976).
45.Sternhell, S., Rev. Pure Appl. Chem., (1964), 14,
   15--46.
46.Evans, M.E., Long, Jr., L., Parrish, F.W., J. Org.
   Chem., (1968), 33, 1074--1076.
47.Edwards, R.G., Hough, L., Richardson, A.C., Tarelli,
   E., Tetrahedron Lett., (1973), 26, 2369--2370.
48.Khan, R., Mufti,K.S.,Parker K.J.,Brit.Pat. 1,430,288
49.Helferich, B., Ber., (1921), 54, 1082--1084.
49a.Helferich, B., Lowa, A., Nippe, W., Riedel, H.,
   Ber., (1923), 56, 1083--1087.
50.Helferich, B., Sprock, G., Bester, E., Ber., (1925),
   58, 886--891.
51.Jennings, H.J., Jones, J.K.N., Can. J. Chem.,(1962),
   40, 1408--1414.
52.Jennings, H.J., Jones, J.K.N., Can J. Chem., (1963),
   41, 1151--1159.
53.Jennings, H.J., Jones, J.K.N., Can J. Chem., (1965),
   43, 2372--2386, 3018--3025.
54.Cottrell, A.G., Buncell, E., Jones, J.K.N., Chem.
   Ind., (London), (1966), 552; Can. J. Chem.,
   (1966), 44, 1483--1491.
55.Khan, R., Carbohyd. Res., (1972), 25, 504--510.
56.Ballard, J.M., Hough, L., Richardson, A.C., Chem.
   Commun., (1972), 1097--1098.
57.Ballard, J.M., Hough, L., Richardson, A.C., Fairclough, P.H., J. Chem. Soc., Perkin 1, (1973),
   1524--1528.
58.Ballard, J.M., Hough, L. and Richardson, A.C.,
   Carbohyd.Res., (1974), 34, 184--188.
59.Khan, R., and Jenner, M.R., (Unpublished results).

Biographic Notes

Riaz Khan, Ph.D., Industrial research chemist. Education at Bristol Univ. Joined staff of Tate & Lyle, Ltd., in 1968. Many papers on carbohydrate reactions and analyses. Philip Lyle Memorial Research Laboratory, Tate & Lyle, Ltd., P.O. Box 68, Reading, Berkshire RG6 2BX, England.

5

Selective Substitution of Sucrose Hydroxyl Groups via Chelates

E. AVELA, S. ASPELUND, B. HOLMBOM, and B. MELANDER
Academy of Finland, Pyorokiventie 10, 00830 Helsinki 83, Finland

H. JALONEN and C. PELTONEN
Abo Academy, 20500 Turko 50, Finland

All of the three primary and five secondary hydroxyl groups of sucrose usually react with the same reagents causing partial derivatization and producing a mixture of degrees of substitution and isomer locations. Due to the multistages of reactions required, the protection of some hydroxyls followed by specific substitutions and reconversions not only causes low yields but usually produces insufficient selectivity. We have found that, if instead, a reaction site in a polyol molecule is converted to a chelate group, usually involving binding two of the hydroxyl groups into a five- or sixmembered ring compound, a high yield selective reaction is obtained (1-4). Such a reaction is, e.g. the selective monoacetylation of methyl 4,6-benzylidene-α-D-glucopyranoside mercuric chelate with acetic anhydride giving a 94 mole-% yield of 2-O-acetyl ester, 2% of 3-O-acetyl and 4% of the di-O-acetyl ester (5). Similarly, the methylation of the cupric chelate of the same compound with methyl iodide gives solely monoethers or 19 mole-% of 2-O-methyl ether and 73% of 3-O-methyl ether (2). The preparations of the selectively reacting chelates are seen in Figure 1.

The substitution reaction can be illustrated by Figure 2. The reason why a bidentate ligand gives solely monosubstitution instead of a diderivative can be explained by blockage of the underivatized oxygen by salt formation as shown in Figure 3.

In the present study, selective substitution via chelates will be applied to etherification and esterification of sucrose. To be a homogenous substrate for a selective reaction, sucrose should give stoichiometrically definable chelates that are sufficiently soluble in a dry aprotic solvent, such as N,N-dimethylformamide or dimethylsulphoxide. One might question whether the unchelated hydroxyl groups of sucrose also would be derivatized.

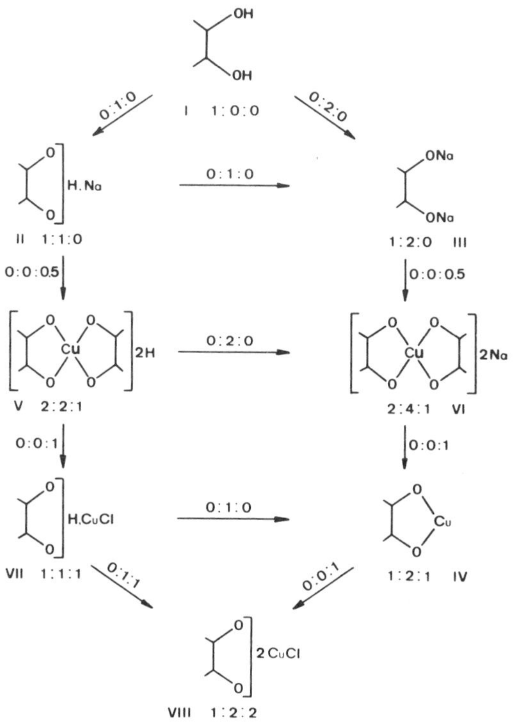

Figure 1. Preparation and interchangeability of mono- and disodium alcoholates 1:1:0 and 1:2:0 resp., and copper (II) derivatives 1:2:1, 2:2:1, 2:4:1, 1:1:1, and 1:2:2 of a dihydroxyl compound. Compounds characterized by mole ratio of diol:NaH:CuCl₂ in synthesis; notations of + and − charges omitted.

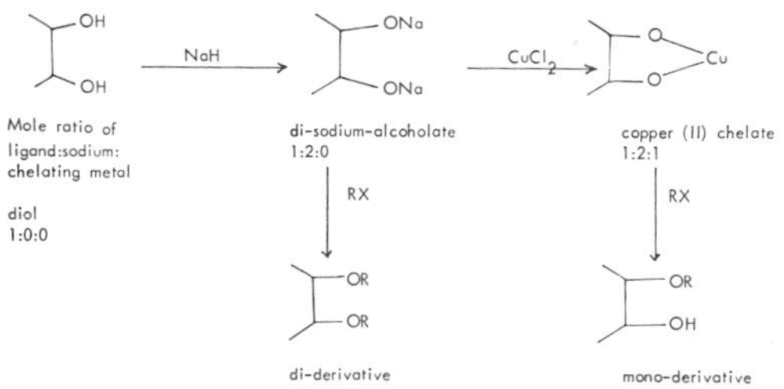

Figure 2

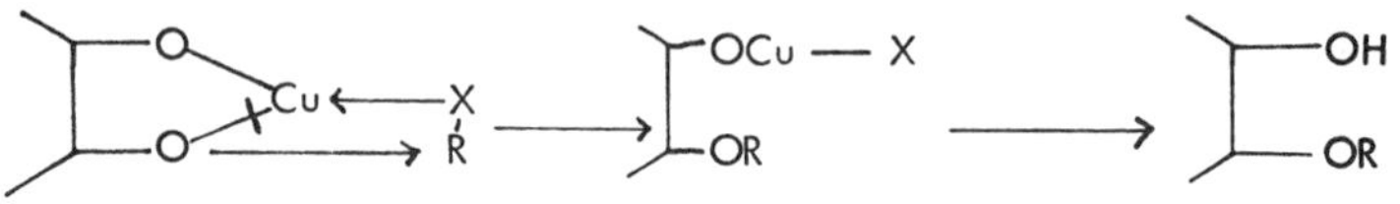

Figure 3

Experimental

Etherification of Sucrose Chelates by Allyl
Halides, and Sodium Bromoacetate. Allyl bromide or
chloride was added to a dimethylsulphoxide (DMSO)
solution of sucrose chelate in the ratios of sucrose:
allyl halide 1:1,1, 1:1,3, 1:2,0 and 1:2,5 and kept at
80°C for 16 to 48 h. The allyl bromide reactions were
carried out in screw cap, sealed test tubes and most of
the allyl chloride reactions in a sealed autoclave.
Decomposition of the sucrose was prevented by keeping
the ratio of sucrose to allyl halide equal or less
than the ratio 1:2,5. The reaction between sucrose
chelates and sodium bromoacetate was performed in the
following ratios; sucrose:bromoacetate, 1:2,6, 1:3,8,
1:5,2 and 1:7,0, in DMSO for 72 h at 70°C.

Esterification of Sucrose Chelates by Acid Anhydrides and Chlorides. Acetic, caprylic, stearic,
maleic or phthalic acid anhydrides were added to
a N,N-dimethylformamide (DMF) solution of a sucrose
chelate in ratios of sucrose:anhydride 1:1,3 or 1:3.
The reaction mixture was allowed to stand with acetic
anhydride for 15 h at 22°C, with caprylic and stearic
anhydride for 3 h at 60°C, or with maleic and phthalic
anhydride for 10 h at 60°C. The isolation, e.g. of
monostearates, was made by extraction with 1-butanol
(6) after addition of 10% aq sodium chloride to the
DMF.
Esterifications with acryloyl, caproyl and lauroyl
chlorides was carried out over 3 h at 60°C. The hydrolysis is avoided when no more than 3 moles of acid
chloride is used per mole of sucrose.

Transesterification. Six to 30 moles of methyl
or ethyl acetate, methyl caprinate, laurate, oleate,
stearate or methacrylate was added to a DMF solution
of one mole of sucrose chelate. The reaction mixtures
were allowed to stand 3 h at 100° or 120°C and analyzed.

Analytical

Sucrose ethers and esters were analyzed as trimethylsilyl derivatives prepared by adding to 0,5 ml of
reaction solution, 0,1 ml hexamethyldisilazane and 0,05
ml dimethylchlorosilane. The carboxymethyl ethers
were gas-chromatographically separated on a 12 ft 3%
OV-17 glass column at first with the temperature programmed at 180°-280^OC, 4°/min, and then isothermally.
The allyl ethers were separated on a 28 m, 0,33 mm id
OV-17 glass capillary column isothermally at 245^OC.
The acetates were gas-chromatographically separated on
a 12 ft 3% OV-17 glass column, with the temperature
programmed at 240° -260° C, 1^O/min, or at 245^OC isothermally. The response values for the partially acetylated sucrose derivatives were determined by interpola-
tion based on values of sucrose and sucrose octaacetate. The caprylates, stearates, maleates and phthalates were separated on a 1 ft 1% OV-1 glass column
temperature programmed at 150°-340^OC, 10°/min. The
acrylates and laurates were separated on a 1 ft 1% SE-
30 glass column with the temperature programmed at
150°-330^OC, 10°/min. The methacrylates were separated
on a 4 ft 3% OV-17, with the temperature programmed at
180°-280^OC, 4°/min. The degree of substitution of the
ethers and esters was determined by LKB gas chromato-
graph-mass spectrometer (GC-MS).
The progress of the esterification reactions was
followed by thin layer chromatography (TLC) in addition
to gas chromatography. Merck silica gel plates were
developed with a mixture of toluene - ethyl acetate -
95 % ethanol (2:1:1 v/v/v) (7). The spots were vis-
ualized by spraying the plates with a solution of 1 g
urea,in 4,5 ml 80% phosphoric acid and 48 ml 1-butanol,
dyed plates heated in an oven at 110^OC for 30 min, and
analyzed with a Vitatron densitometer.

Results

All of the sucrose metal complexes investigated
gave colored solutions except mercury chelates, which
were colorless. Results of the allylation reactions
are shown in Tables I and II. The chelate reactions
are superior in selectivity to the corresponding alcoholate reactions. For instance, in Table II the chelate 2:4:1:4,0 gave 69% mono- and 2% diethers while the
alcoholate gave 50% mono- and 35% diethers. Furthermore, the chelate 2:4:1 was more reactive than the
chelate 1:2:1 and the selectivity decreased for the
chelates when the ratio, sucrose:reagent was increased.

Table I. Composition of the etherification products of sucrose chelates and alcoholates with allyl bromide in DMSO, 80°C, 48 h, mole-%. Reactions performed in screw cap sealed test tubes.

Mole ratio of sucrose: NaH:metal chloride: allyl bromide	Chelating metal	Sucrose			
		Unreacted	Mono-O-ethers	Di-O-ethers	Hydrolysis products
2:4:1:2,6	Co	75	25	-	-
" :4,0	"	59	41	-	-
" :6,0	"	2	-	-	98
1:2:1:1,3	"	89	11	-	-
" :2,0	"	74	26	-	-
1:2:0:1,3	Na	64	29	7	-
" :2,0	"	42	43	15	-

Table II. Composition of the etherification products of sucrose chelates and alcoholates with allyl chloride and allyl bromide in DMSO, 80°C, 16 h, mole-%. Reactions performed in a sealed autoclave.

Mole ratio of sucrose: NaH:metal chloride: allyl reagent	Chelating metal	Sucrose			
		Unreacted	Mono-O-ethers	Di-O-ethers	Hydrolysis products
Allylchloride					
2:4:1:2,2	Co	45	55	-	-
" :4,0	"	29	69	2	-
" :5,0	"	7	-	-	93
1:2:1:2,0	"	63	37	-	-
1:2:0:2,0	Na	15	50	35	-
Allyl bromide					
2:4:1:2,2	Co	63	37	-	-
" :4,0	"	56	43	1	-

The results of the carboxymethylation reaction can be seen in Table III. Here again the chelates were more selective than the alcoholates, although the differences were not as significant as in the former examples. The chelate 2:4:1:7,6 gave 41% mono- and 4% diethers and the alcoholate 22% mono- and 6% diethers. Again, the selectivity of the chelate reaction decreased when greater ratios of sucrose:reagent were used.

Table III. Composition of the etherification products of sucrose chelates and alcoholates with sodium bromoacetate in DMSO, 70°C, 72h, mole-%.

Mole ratio of sucrose: NaH:metal chloride: sodium bromoacetate	Chelating metal	Unreacted	Sucrose Mono-O- ethers	Di-O- ethers	Hydrolysis products
2:4:1:5,2	Co	71	29	-	-
" :7,6	"	55	41	4	-
" :10,4	"	37	50	13	-
" :14,0	"	24	55	21	-
1:2:1:2,6	"	67	33	-	-
" :3,8	"	45	48	7	-
" :5,2	"	23	51	26	-
1:4:2:3,8	"	50	44	6	-
1:2:0:2,6	Na	85	15	-	-
" :3,8	"	72	22	6	-
" :5,2	"	58	32	10	-

The allylation and carboxymethylation reaction velocities differ from each other. For allylation, the chelates reacted slower than the alcoholates, but the carboxymethylation proceeded faster with chelates than with alcoholates. However, the selectivity of the chelates, in this case, is still better than the selectivity of the alcoholates.

The compositions of the product mixtures from the treatment of sucrose with acetic anhydride are given in Tables IV and V. It is seen that, with 1,3 moles of acetic anhydride, the cobalt chelate formed from sucrose, sodium hydride and cobaltous chloride, in the mole ratio 2:4:1, gives 98 mole-% of monoacetates. Even the cobalt chelate 1:2:1 and copper chelates 1:2:1 and 2:4:1 gave over 90 mole-% of monoacetates. The corresponding sucrose alcoholate, 1:2:0, gave no selectivity and at best, 59 mole-% monoacetates. Mass spectrometry disclosed that the peaks 1-8 in the gas chromatogram represent all the possible sucrose mono-

Table IV. Composition of the acetylation products of sucrose chelates and alcoholates with acetic anhydride in DMF, 15h, 22°C, mole-%.

Mole ratio of sucrose: NaH:metal chloride: Ac_2O	Chelating metal	Sucrose		
		Unreacted	Mono-O-acetyl	Di-O-acetyl
1:2:1:1,3	Co (II)	10	90	-
" :3	"	8	92	-
" :1,3	Fe (II)	32	66	2
" :3	"	5	46	49
" :1,3	Cu (II)	4	93	3
" : "	Hg (II)	24	71	5
" : "	Mn (II)	10	89	1
1:2:0:1,3	Na	52	48	-
" :3	"	12	59	20
1:4:2:1,3	Co (II)	29	71	-
" :3	"	1	34	65
" :1,3	Fe (II)	28	72	-
" : "	Cu (II)	4	89	7
" : "	Hg (II)	43	55	2
" : "	Mn (II)	37	63	-
1:4:0:1,3	Na	58	42	-
" :3	"	43	48	-
1:6:3:1,3	Co (II)	34	66	-
" : "	Fe (II)	29	71	-
" : "	Cu (II)	12	84	4
" : "	Hg (II)	62	38	-
" : "	Mn (II)	59	41	-
1:6:0:3	Na	65	31	4

Table V. Composition of the acetylation products of sucrose chelates with acetic anhydride in DMF, 15 h, 22°C, mole-%.

Chelate: Ac_2O	Chelating metal	Sucrose		
		Unreacted	Mono-O-acetyl	Di-O-acetyl
2:4:1:2,6	Co (II)	2	98	-
" :6	"	-	28	72
" :2,6	Fe (II)	21	76	3
" :6	"	4	55	41
" :1,3	Cu (II)	5	92	3
" :1,3	Hg (II)	17	71	12
" :1,3	Mn (II)	26	71	3

Table VI. Content of the different sucrose mono-O-acetates in acetylation products of sucrose chelates and alcoholates, yield of reaction mole-%.

Mole ratio of sucrose: NaH: metal chloride: Ac$_2$O	Chelating metal	Total mono-O-acetate yield	Peak number of different mono-O-acetates in gas chromatograms mole-%							
			1	2	3	4	5	6	7	8
2:4:1:2,6	Co(II)	98	-	68	-	-	-	4	-	26
1:2:1:1,3	"	92	-	54	-	-	-	5	-	33
1:2:1:3,0	"	90	-	54	-	-	-	5	-	31
1:4:2:1,3	"	71	-	60	-	-	-	1	-	10
1:6:3:1,3	"	66	-	58	-	-	-	1	-	7
2:4:1:2,6	Fe(II)	76	4	27	11	8	-	14	5	7
1:2:1:1,3	"	66	5	15	8	18	-	10	5	5
2:4:1:2,6	Cu(II)	92	24	1	38	13	5	4	2	5
1:2:1:1,3	"	92	26	1	35	15	5	2	2	6
1:4:2:1,3	"	89	16	1	38	14	7	2	3	8
1:6:3:1,3	"	84	11	3	35	18	7	1	3	6
2:4:1:2,6	Hg(II)	70	12	6	22	8	9	8	2	3
1:2:1:1,3	"	71	9	18	15	10	12	-	5	2
2:4:1:2,6	Mn(II)	71	8	16	17	11	12	2	2	3
1:2:1:1,3	"	89	5	40	8	10	13	2	5	6
1:3:1:1,3	Fe(III)	65	6	5	12	18	12	3	3	6
2:3:1:2,6	Cr(III)	63	12	7	22	7	9	3	1	2
1:2:0:3,0	Na	57	14	5	11	11	6	2	4	4
1:4:0:3,0	"	48	5	-	17	22	2	-	1	-
1:6:0:3,0	"	31	2	-	13	14	1	-	-	1

acetate isomers. The cobaltous chelates gave only
three different monoacetate isomers, the ferrous chelates gave seven, while all the other chelates studied
Cu (II), Hg (II), Mn (II), Fe (III) and Cr (III), gave
all eight isomers.
     With fatty acid anhydrides, Table VII, the sucrose
chelates gave much higher yields, of monoesters, than
with the corresponding fatty acid chlorides, as shown
in Table IX. The monoester yield also was higher
when, instead of disodium alcoholates, the chelates
were treated with anhydrides. The cobalt chelate
2:4:1 gave 82 mole-% of monostearate or 89% of monoca-
prylate. Even the slower reacting copper 2:4:1 chelate gave a higher and more selective yield than disodium sucrate.
     The copper chelate 2:4:1 gave the same three mono-
stearates but a higher yield than the disodium sucrate.
Cobalt chelate 2:4:1 gave only two monostearates which
based on TLC and GC-MS are not the same ones which are
formed from sodium sucrate.
     The cobalt chelate 2:4:1 reacted with maleic or
phthalic anhydride selectively giving resp. 72 and 85
mole % of monoesters as seen in Table VIII. In contrast the disodium alcoholate gave up to 65% of monoesters. No selectivity was obtained with succinic
anhydride.
     Reacting the fatty acid chlorides with chelates
and alcoholates usually gave primarily monoesters, but
the yield obtained from reacting chelates with anhydrides was lower, as seen in Table IX.

Table VII. Composition of esterification products of
sucrose chelates and alcoholates with caprylic or
stearic anhydride in DMF, 3 h, 60°C, mole-%.

Chelate: anhydride	Chelating metal	Anhydride	Sucrose		
			Unreacted	Mono- esters	Di- esters	Tri- esters
1:2:1:1.3	Co (II)	Caprylic	19	73	8	-
2:4:1:2.6	"	"	-	89	11	-
1:2:0:1.3	Na	"	22	44	28	6
1:2:1:1.3	Co (II)	Stearic	14	61	25	-
2:4:1:2.6	"	"	5	82	13	-
1:2:1:1.3	Cu (II)	"	23	63	14	-
2:4:1:2.6	"	"	25	62	13	-
1:2:0:1.3	Na	"	33	40	27	-

Table VIII.   Composition of esterification products of
sucrose chelates and alcoholates with maleic and phthalic anhydride in DMF, 3 h, 60°C, mole-%.

Chelate: anhydride	Chelating metal	Anhydride	Sucrose		
			Unreacted	Mono-half-esters	Di-half-esters
2:4:1: 4	Co (II)	Maleic	46	54	-
2:4:1:16	"	"	18	72	10
1:2:0: 2	Na	"	31	60	9
2:4:1: 4	Co (II)	Phthalic	90	10	-
2:4:1:16	"	"	15	85	-
1:2:0: 8	Na	"	34	65	1

Table IX.   Composition of esterification products of
sucrose chelates and alcoholates with acryloyl, lauroyl
and oleoyl chlorides in DMF, 3 h, 60°C, mole-%.

Chelate: acid chloride	Chelating metal	Acid chloride	Sucrose		
			Unreacted	Mono-ester	Di-ester
1:2:1:1.5	Co (II)	Acryloyl	49	51	-
2:4:1:3	"	"	42	58	-
1:2:0:1.5	Na	"	44	55	-
1:2:1:2	Co (II)	Lauroyl	46	49	5
2:4:1:4	"	"	26	59	15
1:2:0:1.5	Na	"	39	57	4
1:2:1:1.5	Co (II)	Oleoyl	59	41	-
2:4:1:2.6	"	"	69	31	-
1:2:0:1.5	Na	"	92	8	-

Increasing the amounts of acyl chloride relative to sucrose to molar ratios of over 3:1 did not improve the ester yield but rather initiated the hydrolysis of sucrose. Hydrolysis was not found to occur at lower levels of acyl chloride. A ratio of sucrose to alkyl chloride of 1:1.5 was found to be preferable. A similar reagent threshold amount was found in the etherification of sucrose with alkyl halides. The use of 1.3 moles of e.g. methyl iodide per mole of sucrose chelate seemed preferable. Ratios over 3 moles of alkyl halide caused sucrose hydrolysis, as was observed with earlier attempts at sucrose esterification (8).

The transesterification of sucrose with the readily available, fatty acid methyl esters in the presence of alkaline catalysts (6,9) is a common method of preparing sucrose fatty acid esters with a low degree of substitution. When the reaction was applied to sucrose chelates, it produced predominantly monoesters, with diesters. The chelate transesterification thus is more selective and gives higher yields than the standard transesterification. For example, the sucrose cobalt chelate 2:4:1 with ethyl acetate yielded 92 mole-% of esters with the composition of 68% of monoacetate, 24% of diacetate and 8% of unreacted sucrose. The chelates regularly gave higher sucrose ester yields (60-92%) than the corresponding disodium stearates (32-63%) in the transesterifications studied. Increasing the reaction temperature from 80°C to 120°C, or the amount of methyl ester, improved the sucrose ester yield.

Reaction of sucrose chelates with acid anhydrides and chlorides gives selective production of monoesters and only some diesters, in contrast to the reaction products of sucrose alcoholates with acid anhydrides, chlorides and esters giving sucrose mono- as well as diesters. This result is in good accordance with our earlier observations of selective substitution of polyols via chelates (1-5). In this case as well, the reason why only one of the two oxygen atoms of a chelate group is derivatized is obviously that further substitution of another oxygen is blocked by salt formation. Adding water will convert this oxygen back to a hydroxyl group, while the other oxygen is derivatized to ester, as shown in Figure 4.

In case of transesterification, the blocking reaction is not effective because a cobaltous dialcoholate, $-O-Co-O-CH_3$ is formed which reacts further causing partial esterification of the other oxygen which was involved originally in the chelate formation. The principal mono- and partial diester formation from

Table X. Composition of transesterification products of sucrose chelates and alcoholates with ethyl acetate, methyl caprinate, laurate, stearate, oleate, and ethyl methacrylate in DMF, 3 h, 80°-120^OC, mole-%.

Chelate: ester	Chelating metal	Ester	Unreacted	Mono-ester	Di-ester
2:4:1:20	Co (II)	Acetate	29	45	26
1:2:0:20	Na	"	68	24	8
2:4:1:20	Co (II)	Caprinate	30	58	12
1:2:0:10	Na	"	61	34	5
2:4:1:20	Co (II)	Laurate	35	56	9
1:2:0:15	Na	"	55	38	7
2:4:1:20	Co (II)	Stearate	34	55	11
1:2:0:10	Na	"	37	50	13
2:4:1:20	Co (II)	Oleate	40	52	8
1:2:0:10	Na	"	58	32	10
2:4:1:12	Co (II)	Methacrylate	45	45	10
1:2:0: 6	Na	"	38	37	25

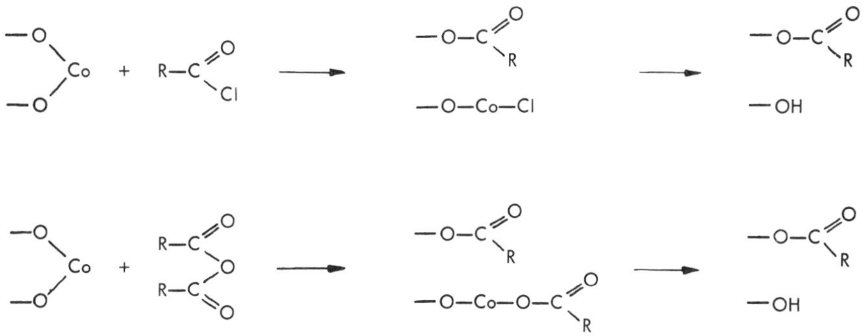

Figure 4

sucrose via transesterification, can be written as
shown in Figure 5.

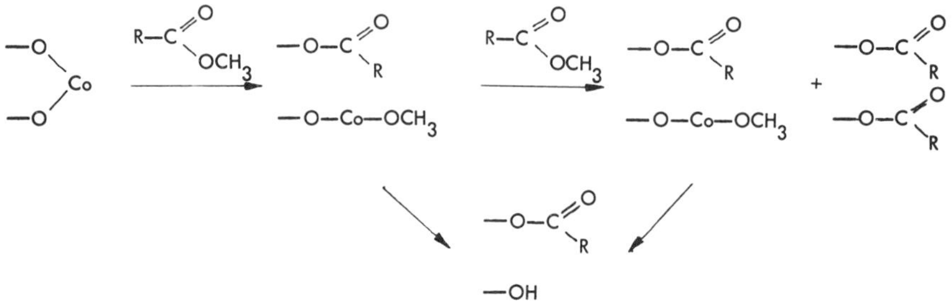

Figure 5

Summary

Sucrose chelates in DMSO react selectively with
allyl halide and sodium bromoacetate to produce monoethers in high yields.   Sucrose chelates in DMF solution react selectively with acid chlorides, anhydrides
and esters to produce sucrose monoesters or, in some
cases, diesters as well.   The yields and selectivities
of the partial esterification of sucrose via chelates
is higher than with other methods.   The yield range
of the acid derivatives studied are shown in Table XI.

Table XI.   Composition of Esterification Products of
sucrose chelates with acid derivatives, mole-%.

Acid derivative reacting with sucrose chelate	Yield of sucrose ester, mole-%		
	Total esters	Mono- esters	Di- esters
Acid chlorides	41-74	41-59	0-15
Acetic acid anhydride	90-98	90-98	0
Higher fatty acid anhydrides	95-100	82-89	11-13
Dicarboxylic acid anhydrides	82-85	72-85	0-10
Acetic acid esters	71-80	43-50	26-30
Methacrylic acid ester	55	45	10
Higher fatty acid esters	60-70	52-58	8-12

Acknowledgements

   The financial aid from The International Sugar Research Foundation, Inc., Bethesda, Maryland, U.S.A. and from The National Research Council of Technical Sciences of The Academy of Finland is gratefully acknowledged.

Abstract

   The partial derivatization of sucrose hydroxyls generally gives isomer mixtures with various degrees of substitution. We have reported earlier a method to selectively monoetherify or monoesterify polyols and glucosides via metal chelates instead of using multistage organic protection and reconversion reactions. Sucrose chelates also give directly a selective and, to a certain DS level, limited reaction. The chelates are prepared in anhydrous DMF or DMSO by ionization of the desired number of hydroxyl groups of the sucrose molecule with stoichiometric amounts of sodium hydride to form alcoholates which, with metal salts, give the chelates. The etherification of sucrose with alkyl halides or esterification with organic acids causes hydrolysis. The hydrolysis or diether formation is avoided if sucrose chelate is etherified at moderate temperatures and with only a small excess of allyl halide or sodium bromoacetate, giving 55-69% mono- and 0-2% diallyl ethers respectively, 41-48% mono- and 4-7% dicarboxymethyl ethers of sucrose. The partial esterification of sucrose with organic acid chlorides, anhydrides or esters gives mixtures of esters and a low yield. Acid chlorides with chelates in anh. DMF give monoester 41-59 mole-%, diester 0-15% and no higher esters. Acid anhydrides give mainly monoesters, low molecular weight anhydrides 80-98%,and higher fatty acid anhydrides 82-89%.Transesterification of chelates with acetic acid ethyl ester gives monoesters 45% and diesters 26%, and higher molecular fatty acid esters give monoester 52-58% and diester 8-12%.

Literature Cited

1. Avela, E., Finn. Appl. 2578/1971, Brit. Pat. 1 410 033/1975, Austrian 321319/1975, Can. 981663/1976, U.S. 3,972,868/1976.
2. Avela, E., and Holmbom, B., Acta Acad. Aboensis, (1971), 31, 14, 1.
3. Avela, E., Melander, B., and Holmbom, B., Ibid

   (1971), 31, 15, 1.
4. Avela, E., La Sucrerie Belge, (1973), 92, 337.
5. Avela, E., and Melander, B., Paper presented at VI
   Int. Symp. on Carbohydr. Chem., 1972, Madison,
   Wisc., U.S.A.
6. Hickson, J.L., Sucrose Ester Surfactants, Int.
   Sugar Res. Found., Inc., New York 1960, pp. 1-10.
7. Weiss, T. J., Brown, M., Zeringue, Jr., H.J., and
   Feuge, R.O., J. Am. Oil Chem. Soc. (1971), 48,145
8. Black, W.A.P., Dewar, E.T., Paterson, J.C., and
   Rutherford, D., J. Appl. Chem. (1959), 9, 256.
9. Osipow,L., Snell, F.D., York, W.C., and Finchler,A.,
   Ind. Eng. Chem. (1956), 48, 1459.
10.Avela, E., Abstracts of papers, 138th ACS Nat.
   Meeting, New York, (1960), 18D, No. 51.

Biographic Notes

     Professor Eero Avela, Ph.D., Prof. of Chem.
Educated at Helsinki Univ., Oxford Univ., State Univ.
of New York, and Hamburg Univ.   Director of Research
at Oy Kaukas Ab (Finland), Norland Papier GmbH & Co.
(West Germany), Dir. of Res., Academy of Finland,
Prof. of Technical Polymers and Plastics Chemistry,
Abo Academy, Finland.   Some 86 papers and 4 patents
on cellulose, polymers and carbohydrates.   Academy of
Finland, Pyorokiventie 10, 00830 Helsinki, 83 Finland.

Discussion

Question: Has there been any study of toxicity of, or LD_{50} work done with, the tetrachloro"galacto" sucrose derivative?

Professor Hough: Studies have been initiated by Tate and Lyle but, as far as I know, the results are not yet available.

One must note, of course, that we came across the sweetness accidentally. It obviously is a hazardous process to taste compounds haphazardly. Yet, because of this accidental discovery, what we have to do now in our program of work, which we have not done previously because we never suspected that these compounds would be sweeter than sucrose, is to submit all of our compounds to toxicity screening before we taste them. Otherwise, I might lose many research students in the process, and they are too valuable to lose!

I might just comment on how unpredictable the sweetness was, because it has been believed that all the monoacetates are less sweet than sucrose. Sucrose octaacetate, as I mentioned, is very, very bitter. The general prediction, therefore, was that you cannot make sucrose sweeter than it is.

Question: Although one may not find toxicity in such a test, it might be that the compound may be a competitive inhibitor, which could result, for example, in hypoglycemic activity. There is a further complication; it could be hydrolized, and probably would be hydrolized to chlorogalactose and chlorofructose. If so, someone has to study a great deal of the biochemistry of chloroglucose and chlorofructose. As far as I know, no studies have been carried out in this regard.

Professor Hough: Dr. Norman Taylor, of course, is
quite correct. As an expert on fluorocarbohydrates, he
is well aware of the problems in this field.   He has
made it a lifetime study.   I shall hope that he will
take up chlorocarbohydrates now, and assist us in our
work in exploiting the chloroderivatives of sucrose by
studying their biochemistry.

Professor Taylor: May I add that saccharin is an
inhibitor of glucose transport.

Question: Are the dianhydrosucrose derivatives
appreciably soluble in organic solvents?

Professor Hough: Yes, they are.   They are more
soluble than sucrose itself, which, for example, is
difficult to dissolve in pyridine.   The dianhydrosucrose dissolves very much more readily.

Question: Would this be a route to further derivatization?

Professor Hough: Certainly, yes.   It is a very
important route to further derivatization in my view
because only four hydroxyls are now exposed, which is
something that we can exploit, and the dianhydride is
readily available.

Question: Is it possible to prepare ^{14}C-labelled-
galacto-sucrose?

Professor Hough: Dr. Riaz Khan could make expert
comments since he has described a good synthesis of
galacto-sucrose in Carbohydrate Research.   Taking ran-
domly-^{14}C labelled sucrose, presumably prepared photo-
synthetically, it could be converted readily into
galacto-sucrose.   The important objective is a good
radiochemical yield.

Dr. Khan: The synthesis of galacto-sucrose from
sucrose proceeds in quite a good yield.

Question: Why are the bulky sulphonylchlorides
more selective for the C-1^1-position?

Professor Hough: We would have predicted that reaction would occur preferentially at the 6,6^1-posi-
tions.   This, in fact, does happen if you reduce the
quantity of the reagents.   The next reactive position
certainly is the 1^1-position in sulphonylation.   This

reaction does depend, I believe, upon the relative ac-
cessability and acidity of the hydroxyl groups, in contrast to nucleophilic substitution reactions. When
you introduce a fourth sulphonate substituent, it goes
to the C-2 grouping of sucrose, similar to etherification. I think this accounts for the reactivity of the
1^1-position.
     If you look closely at the mechanism of the reac-
tion, it depends largely upon the stereochemistry,
which is all right at the primary positions. In addition, the hydroxyl group must react with base and the
sulphonylhalide with the removal of hydrogen, or a proton, from that primary position. This is more favorable at the 1^1-position than it is, for example, at the
4-position. This is how I would explain the reactivity.

     Question: In bimolecular nucleophilic substitu-
tion reactions, the 1^1-position is very hindered.
Would you comment?

     Professor Hough: Nucleophilic substitution is
quite different in mechanism from a selective acylation
reaction because, in the latter, only the hydrogen of
a hydroxy group is being replaced by a bulky group.
No attempt is being made to replace oxygen plus the
bulky group by a rearward attack. Therefore, esterification at the 1^1-position will proceed as at a normal
primary hydroxyl.

     Question: Dr. Khan, would you repeat the condi-
tions you used for your hypooxidations? You had high
yields, and you do not lose any of the acetate groups.

     Dr. Khan: Actually, the product is completely
deesterified but, in order to isolate it, we reesteri-
fy it, and then we get the acetyl derivatives.

     Question: Professor Hall, have you established
which protons in the fructose residue are responsible
for the relaxation of ^{1}H?

     Professor Hall: I am afraid not. To do this, we
need either a 1,000 megacycle n.m.r. machine, which is
not available, or to make the corresponding deuterated
derivatives. We would be very interested to know if
anybody has any good ideas on how to get the carboxylic
acids, because then we could get the deuterium in.
That is the reason why we would like to get deuterated
sucrose derivatives.

Surfactant Esters

Introduction

A. J. VLITOS

Tate & Lyle Ltd., Group R & D, P.O. Box 68, Reading RG6 2BX, Berks, England

The topic of the first session of this Symposium
was some of the basic discoveries, and some of the more
general concepts surrounding sucrochemistry.   This,
second session, deals with applications. Specifically,
there are three papers dealing with surfactants, two
relating to esters and one paper on organometallic
derivatives.

It is well to remember that it was Dr. Hass who
originally suggested (in 1952) that surfactant esters
might be produced from sugar.   As a direct result, the
Foundation programmes at Foster D. Snell culminated in
the granting of patents as well as licences in Japan
and France.

Surfactants are used in specialty applications in
foods, animal feeds and pharmaceuticals - not only in
washing detergent powders as many people believe!   In
recent years, two sugar companies themselves have prov-
en that surfactants produced from sugar indeed are
feasible.   Raffinerie Tirlemontoise in Belgium and
Tate & Lyle Ltd., in England have led the way to im-·
proved routes for synthesis.

The sugar alcohol studies in Holland at Chemie
Combinatie and the organometallic studies by Poller at
Queen Elizabeth College represent ester types of derivatives also, although aimed at entirely different markets.

The future of sucrochemistry will depend on producing marketable products which are competitive with
existing ones.   It is a future wholly dependent upon
the correct combination of technical skill and marketing expertise - and well within the realm of probabili-
ty!

6

New Plant and New Applications of Sucrose Esters

T. KOSAKA and T. YAMADA

Ryoto Co. Ltd., 5-2, Marunouchi 2-chome, Chiyoda-ku Tokyo 100, Japan

It was 15 years ago that sucrose ester (SE) made
a spectacular debut in the industrial world as one example of sucrochemistry.   In 1960, Ryoto's predeces-
sor,   Dai-Nippon Sugar Mfg. Co., Ltd., built a semicom-
mercial plant with an output of 300 tons per year.
Since then, there has been a growing demand for SE,
mainly in the food industry, as a very safe product
having as its raw materials sugar and natural edible
fats, and as a near-natural surfactant.   In 1967,
there was completed a full-scale plant with a continuous process scaled-up to 1200 tons per year.   This
process, known as the Hass-Snell process,used dimethylformamide as the solvent for a transesterification of
sucrose by the methyl esters of fatty acids to yield SE.
In 1967 Dr. Osipow and coworkers developed the
Nebraska-Snell process, in which a microemulsion of
sucrose is formed in a propylene glycol solvent and
treated with methyl esters of fatty acids.   This method
was then improved in Japan by Daiichi Kogyo Seiyaku
Co., Ltd., who succeeded in industrializing it by using
water instead of propylene glycol.
Other methods of sucrose ester synthesis include
a nonsolvent method known as the USDA method, for which
Ryoto is exclusive licensee, and the Zimmer method developed in West Germany.   Industrialization of these
methods has been studied by at least 10 companies
throughout the world but, while some may still be continuing their research, the only companies among them
to have commenced commercial production of SE are Ryoto
and Daiichi Kogyo Seiyaku in Japan.   As a sucroglyceride producer, there is Rhône-Poulenc of France and recently Tate & Lyle, Ltd., of the U.K. has developed an
SE-detergent (discussion of both occur later in this
volume).   It is rather puzzling why SE industrialization should have occurred only in Japan, which is

totally dependent on imports of the main raw materials, sugar and edible fats and whose per capita consumption of emulsifiers is, in fact, lower than that of Europe or of the U.S.A.

In 1975, Ryoto completed and put into operation a new plant, producing 3000 tons per year (including some SE compound products) to cope with the growing domestic and overseas demands.

I am sorry to say that, since it is Ryoto's policy the new process of this plant is not to be disclosed for some time yet, I shall go no further here than to say that it is the most advanced process available today, both from the angles of public health and economics, being the crystallization of our 15 years of technical experience with SE.

In particular, from the product safety view point, although residues of solvents, especially dimethyl-formamide (DMF), often have been called into question even when within the food additive standards, the product of this process is solvent-free and will cause no problems, whatsoever. FAO/WHO had approved a maximum level of 50 ppm for DMF, but when revaluation of the SE standards was carried out in April of this year, we requested FAO/WHO to revise the DMF level to, "not detected". The requested level was accepted. In all other specifications, this product meets not only the Japanese but also all foreign food additive standards. Furthermore, the product naturally passes the standards of not more than 10 ppm methanol and not less than 90% SE content, implemented by a directive of the EEC Committee on June 18, 1976.

Table I
Current Usage of Sugar Esters in Different Applications

Cakes and Breads	31.0%
Emulsified Oil and Fat	22.3%
Coffee Whitener, Whipped Cream	
Recombined Milk, Shortening Oil,	
Ice Cream	
Instant Food	13.1%
Curry, Soy Bean Curd, Cocoa,	
Cake-Mix	
Confectionary	15.3%
Biscuit, Chocolate, Chewing-gum,	
Rice-Cake, Tablet Candy	
Detergent	9.3%
Others	9.0%
Drugs, Cosmetics,	
Chemical Industries	

New Applications

Many applications of SE in the food field already
have been discussed.   As much as 80% of the SE demand,
so far, occurs in the food industries, in applications
as shown in Table I and Table II.

Table II

Representative Types of Ryoto Sugar Esters and their
Uses.

Types	HLB	Main Fatty Acid	Representative Uses
S-370	2-3	Stearic	Tablet Lubricant Shortening, Margarin Emulsifier
S-570	5	Stearic	Chewing-gum Emulsifier Chocolate Viscosity Reductioner Instant Curry Emulsi- fier
S-770	7	Stearic	Biscuit Emulsifier, Improver Rice Cake Improver Caramel, Candy Emulsi- fier
S1170	11	Stearic	Ice Cream Emulsifier Cake Foaming Agent Milk, Dairy Product Stabilizer
P1570	14-15	Palmitic	Oil, Fat Emulsifier Instant Food Wetting Agent Insoluble Material Dispersing Agent
LW 1540	15	Lauric	Detergent Surfactant
OW 1540	15	Oleic	

However, we expect that, from now on, SE demand will
occur in fields other than food.   As an example of
SE's potential, I would like today to discuss its
application in detergents.
     The research on SE was originally motivated by the
need for development of a nonirritant, nontoxic detergent.
     Much research was done on SE detergent development
with the objective of preventing environmental pollu-

tion and damage to human health. However, this research was, and is still, handicapped by the fact that,
for SE itself the detergency defined as its (grease-
removing power), cannot compete with that of synthetic
detergent materials designed primarily for efficiency.
On the other hand, the property of being a safe detergent must surely find great emphasis when one wishes to
clean foodstuffs and food materials, whether in the
kitchen or in factory processing. In view of the
penetration of detergent components into foodstuffs
and food materials, in addition to the residues left
by inadequate rinsing, an ideal in the future would be
for all foodstuff detergents to be made of natural
products and/or food additives.
    For cleansing foodstuffs, the detergency and
strong grease-removing power of the synthetic detergents is unnecessary. We have carried out experiments
on virtually every combination of food additives which
could be used as detergent components, and have fabricated a product which actually surpasses synthetic
detergents with respect to the effects which we sought.
    Choosing from the safety viewpoint, the following
substances were selected as surfactant, builder and
solvent for these tests.

Table III

Surfactants:

sucrose esters: 16 types   stearic, oleic
                                 lauric, palmitic etc.
monoglycerides: 3 types   capric, lauric, oleic
sorbitan esters: 4 types   lauric, stearic, pal-
                                   mitic, oleic
(polysorbates: 2 types   lauric, oleic)

Builders:

Na-maleate,   Na-succinate,   Na-glutamate
Na-lactate,   Na-oxalate,   Na-citrate
Na-gluconate, K-pyrophosphate

Solvents:

glycerol, propylene glycol, ethanol

    The resulting detergent is comprised of the in-
gredients shown in Table IV and exhibits the characteristics shown in Table V.

Table IV
Composition of SE Detergent

1. Sucrose ester of fatty acid   (Food additive)
   Sucrose cocoate
   Sucrose tallowate
2. Sodium citrate   (Food additive)
3. Propylene glycol   (Food additive)
4. Ethanol   (Food grade)
5. Water

Table V

Characteristics of SE Detergent

1. Free from any fear of toxicity, because it consists of food additives.
2. No trouble to the skin such as chapped hands due
   to the detergent.   Rather, sucrose ester
   protects the skin.
3. Excellent in cleansing dirt, agricultural chemicals and bacteria adhering to vegetables,
   fruits, etc.
4. No fear of environmental pollution, because it is
   capable of complete biodegradation.
5. Inhibitory action on the growth of bacteria.
6. Less residues than synthetic detergents on food
   materials and foodstuffs.
7. Promotion of operation efficiency because foaming in cleansing is slight.

The back-up data for my statements of these
characteristics is explained in outline form in Table
VI-XV, Figure 1, and Table XVI.
The uses in foodstuff cleansing naturally are
expected to be centered on the food processing industry.
Such uses include cleansing of raw materials for
fruit and vegetable juices, including mandarin oranges
and tomatoes, as well as of marine products, poultry,
and frozen food materials, etc.   They also include
cleansing of the ingredients used in cooking in the
service industries, such as restaurants, hotels and in
hospital catering.
In the domestic kitchen, since the detergent will
also be used for dish-washing, its grease-removal
power might become a problem, for unfortunately the
results of dish-washing tests did not come up to synthetic detergent performance.   The question of whether
greater weight will come to be placed on grease-removal

efficiency or on safety, including that of the environment, that is, what we might call the level of public
awareness of safety, will most likely be a key factor
effecting the adoption of this detergent for household
kitchen use.

Table VI

Safety

1. Acute Toxicity LD_{50}
    Oral administration
    a group of 10 male mice, 33.96 ml (38.23 g)/kg
    a group of 10 female mice,24.8 ml (27.91 g)/kg

    Reference: Synthetic kitchen detergent LD_{50} 6-10g/
        kg, Japan Association of Synthetic
        Detergents for Households: Problems on
        Synthetic Detergents, November, 1972.

Table VII

Safety

2. TLm   on Killifish

    a) Test Conditions:
    Killifish;   average length   3.91 cm
                      average weight   0.47 g
    Test water temperature;   2.5+ 1°C
    Diluted water;
        pH 7.0, alkalinity 0.4 meq/l
        hardness   25 ppm with inorg. salt
        1 l. of test water to 1 gram of fish weight
    b) Test Method: Japanese Industry Standards
    c) Test Results:

	24 hr	48 hr
SE Detergent	410	400
Synthetic Kitchen Detergent (LAS)	48	48
Synthetic Kitchen Detergent (Fatty Alcohol)	42	38

Unit:ppm

Table VII, Safety, continued

3. Biodegradation of Sucrose Esters

	Biodegradation
Sucrose laurate	100% 1*
Sucrose tallowate	100% 2*

*1 H.J. Heinz, W.F. Fischer: Fette-Seifen-Austrich-
   mittel, 69 (3), 188-196 (1967)

*2 C.H. Waynan, J.B. Robertson: Biotechnol. Bioeng.,
   5, 367-384 (1963)

Table VIII

Test on Skin Irritation

1. Rabbits

    a)   Test rabbits:

    Four each male and female,
    Japanese white species weighing about 3 kg

    b)   Method:

    Hair on the left side of the back was shaven.
    Two application areas, 10 cm^2 (3.3 x 3.3).
    One area was treated with distilled water as
    control. The other with 0.5 ml sample of the
    Detergent.

    c)   Application period:

    One week (once a day).

    d)   Test results:

    No development of erythema, edema, crust etc.,
    or abnormality of pyrexia etc., was found.
    No difference was found between control and
    sample throughout the entire test period.

Table IX

Test on Skin Irritation

2. Human Body
    Effectiveness of Sucrose Cocoate as an Irritation
    Mitigator for Alkylbenzenesulfonate.
    a)   Test method: Patch Test
          15 persons, 30% concentration, 2-16 hr
    b)   Test results:

Sodium alkyl benzene- sulfonate (%)	Sucrose coconut oil ester (%)	Persons with vigorous Reaction	Persons with reaction	Persons without reaction
30	0	7	7	1
29	1	4	10	1
28	2	1	12	2
27	3	0	12	3
20	10	0	6	9
15	15	0	3	12
0	30	0	0	15

Table X

Cleansing Tests

1. Dirt: Semidry Dirtied Cotton Clothes
    - Composition of dirt, rubbed into the clothes

       Soil   49.0 wt %   n-Decane   5.0 wt %
       Carbon black   0.5   Ferric oxide   0.5
       Beef tallow   30.0   Liquid paraffin   10.0
       Cetyl alcohol   5.0

    - Cleansing and Rinse: by Terg-O-Tometer
    - Measurement: by Reflectometer

       Cleansing   Reflexibility of   Reflexibility of
       Efficiency % =   Dirtied cloth   -   Dirtied cloth
                         after cleansing   before cleansing   x 100

                         Reflexibility of   Reflexibility of
                         original cloth   -   Dirtied cloth
                                                 before cleansing

    - Test results

SE Detergent	Synthetic Kitchen Detergent	Water
39.5	30.0	10.0

Table XI

Cleansing Tests

2. Agricultural Chemicals

- Commercially available vegetables and fruits were spontaneously dried for one day after being sprayed with agricultural chemicals.

- Cleansed with propeller type agitator

- Cleansing rate (%)

$$\frac{\text{Amount of agricultural chemicals in the cleansing liquid}}{\begin{array}{l}\text{Amount of agricultural} \\ \text{chemicals in the} \\ \text{cleansing liquid}\end{array} + \begin{array}{l}\text{Amount of residual} \\ \text{agricultural chemicals} \\ \text{on vegetable}\end{array}} \text{X 100}$$

- Vegetables

  Tomato, Cucumber, Cabbage, Spanish paprika, Apple, Orange

Table XII

Test results: Cleansing rate of Agricultural Chemicals (%)

	TOMATO		CUCUMBER		CABBAGE	
	Zineb	Vordo	Chloro phalonil	Zineb	Vordo	Fenitro-phion
Water	11.5	0	65.0	19.5	38.0	58.5
SE Detergent	64.0	73.5	97.5	62.0	90.5	79.0
Synthetic Kitchen Detergent	45.0	7.5	70.0	69.0	63.0	77.5

	Spanish Paprika	Apple		Orange	
	Chlorophalonil	Zineb	Chloro-phalonil	Zineb	Vordo
Water	60.0	84.0	50.0	12.5	0
SE Detergent	93.5	91.0	91.0	52.5	93.5
Synethetic Kitchen Detergent	91.0	90.0	92.0	28.0	3.5

Analysis: Zineb ... as Zn by atomic absorbtion analysis

Vordo ... as Cu   "

Chlorophalonil ... by ECD gas chromatography

Fenitrophion   ... by FPD   "

Table XIII

Cleansing Tests

3. Bacteria
   - Commercially available vegetables
     (Number of bacteria are between 10^5 and 10^6 per g usually)
   - Cleansing by shaking for 3 min
   - Test results:

Bacteria Removal Rate (%)

	Cucumber		Cabbage		Spanish Paprika	
	Common Bacteria	E. Coli	Common Bacteria	E. Coli	Common Bacteria	E. Coli
Water	60.5	48.0	64.5	70.5	67.5	67.5
SE Detergent	98.5	96.5	91.5	89.0	97.5	91.5
Synthetic Kitchen Detergent	95.5	94.5	94.0	87.0	-	-

Calculated by the same way with that of Agricultural Chemicals

Noted for Cleansing Tests:

The concentration of each detergent in cleansing tests was due to the direction of standard use:

SE Detergent: 0.25%   Synthetic Kitchen Detergent: 0.17%

Table XIV

Residual Amounts of Surfactant for Detergent on Vegetables*

- Preparing of labelled surfactant
  Sucrose ester: alcoholysis of 14_C labelled fatty acid and sucrose

  ABS: synthesis of 25_S labelled H_2SO_4 and alkyl-benzene

- Test conditions
  Vegetables were dipped for 5 minutes, agitated at certain intervals in surfactant solution. Rinsed for 1 minuted with 100 ml water per 4 x 4 cm of surface area of vegetable.

- Test results: mg/100 g

	Sucrose ester		A B S	
Concentration	0.1 %	0.2 %	0.1 %	0.2 %
Cabbage	0.015	0.019	0.19	0.28
Cucumber	0.0011	0.0015	0.07	0.08
Radish	0.0007	0.001	0.018	0.027
Tomato	0.0007	0.0011	0.01	0.019
Grape	0.001	0.0017	0.037	0.07

* (Cited from the technial report of Kao Soap Co., Ltd.)

Table XV

Inhibitory Action on Growth of Bacteria

1. Influence on Growth of E. Coli

Concent- ration mg/ml	550 mμ absorbance (-log T)		
	4 hr	6 hr	8 hr
Sucrose laurate	1.0	0.050	0.050	0.050
	0.1	0.015	0.010	0.010
Lauric acid	1.0	0.072	0.062	0.105
	0.1	0.160	0.550	0.630
Methyl laurate	1.0	0.162	0.499	0.660
	0.1	0.185	0.610	0.770
Tween 20	1.0	0.150	0.520	0.600
	0.1	0.128	0.480	0.630
Nonaddition	-	0.160	0.565	0.650

Medium: $Na_2HPO_4 \cdot 2H_2O$ 8.8, KH_2PO_4 3.0,

NH_4Cl 1.0, $MgSO_4 \cdot 7H_2O$ 0.02,

$FeSO_4 \circ 7H_2O$ 0.005,   Glucose 4.0

Cultured at $37°C$

Biochemical and Biophysical Research Communications

Taken from Kato, A., Arima, K., Biochem. Biophys. Res. Commun. *42 (4), 596–601.*

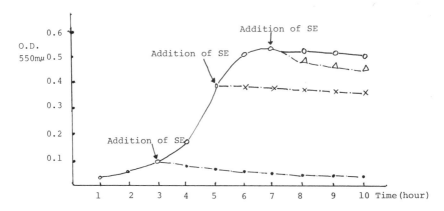

Biochemical and Biophysical Research Communications

Figure 1.   Inhibitory action of growth of bacteria. Influence of addition of sucrose laurate on various reproduction conditions of E. Coli *(Taken from Kato, A., Arima, K.,* Biochem. Biophys. Res. Commun. *42 (4), 596–601.*

Table XVI

Comparison of Foaming Power

By modified method of Weeks

			ml: Foaming Volume
	SE Detergent	Soap	Synthetic Kitchen Detergent
Concentration (%)	0.25	0.75	0.17
Deionized water	418	6,315	1,385
Water of 40 ppm CaCO$_3$	318	215	1,389
City water	140	33	1,213

A further point worth special mention is the recent marketing of this detergent for washing babies' nursing bottles. It has been found to be superior to synthetic detergents in removing milk residue from the bottle surface, as Table XVII shows. Of course, it meets the social demand for the safety of the nursing child.

Table XVII

Cleansing Test

Test on Milk Dirtied Plates

- Test method:
  Dried for 3 hr, 80°C, after being applied with commercially available milk powder solution on plates and left it as it was. Dipped for 6 hr in detergent liquid.
- Test results: Cleansing rate (%)

	Detergent Concentration	Glass plate	Polycarbonate plate
SE Detergent	0.15 0.25	93.5 98.6	86.5 95.6
Water	–	75.4	70.9
Synthetic Kitchen Detergent (LAS)	0.17	78.0	–
Synthetic Kitchen Detergent (Fatty alcohol)	0.17	76.8	–

I am convinced that SE detergent gradually will
spread not only for special uses like the above-men-
tioned ones but also, from the viewpoint of safety, for
wide use in household kitchens.
At present, based on the promising applications of
SE detergency to specialty detergents, we now are undertaking research and development work on applications
of SE in shampoo and toothpaste.   I trust to have the
opportunity, in the near future, to report on the results of this work.

Abstract

      Since Dai-Nippon Sugar Mfg. Co. (Predecessor
of Ryoto Co.) constructed the first commercial plant
with productivity of 1,200 ton/year in 1967, the .de-
mand of sucrose esters has increased 15-20% per annum
and reached a 1,000 ton/year total in Japan, Asia and
Europe.   In order to meet demand which is expected to
grow, Ryoto completed a new plant capable of producing
3,000 ton/year at the end of 1974.   The products are
completely solvent-free.   The new process developed
by Ryoto does not employ dimethylformamide as a solvent which has often caused public discussion about
the safety of sucrose esters.   The quality of the new
products meets the additional requirement of EEC common approval that the total sucrose ester content
shall be not less than 90% and the total methanol
content (free and combined) shall be not more than 10
mg per kg, and the specifications of FAO/WHO.   Sucrose esters are of great promise not only in food
fields but also in non-food uses, especially in detergents.   Detergents made with sucrose ester, sodium
citrate, propylene glycol and ethanol have been expanding the market for food and household kitchen purposes due to edibility, mild washing ability and 100%
biodegradation.

Biographic Notes

      Terahiko Kosaka, Dir. Devel. Dept.   Educated at
Faculty of Agric. Chem., Tokyo Univ.   Joined Dai-Nip-
pon Sugar Mfg. Co., Ltd. in 1952.   In 1973 became Dir.
Devel. Dept., specializing in yeast and sugar ester
surfactants.   Ryoto Co.,Ltd., 5-2 Marunouchi, 2 Chome,
Chiyoda-Ku, Tokyo, Japan.

7

Sucrose Ester Surfactants—A Solventless Process and the Products Thereof

KENNETH J. PARKER, K. JAMES, and J. HURFORD

Tate & Lyle Ltd., Group R & D, Philip Lyle Memorial Laboratory,
The University, Whiteknights, P.O. Box 68, Reading, Berks, RG6 2BX, England

Sucrose is unique in its combination of physical and chemical properties with ready availability. As a non-reducing sugar, it is extremely stable except to hydrolysis, and a strongly polar hydrogen-bonding structure results in its very high solubility in water. It is the lowest cost, polyhydric alcohol available, and its world production exceeds that of any other single, pure organic chemical.

It is not surprising, therefore, that the industrial potential of sucrose derivatives of long chain alkyls (1) as surface-active agents has been long recognised. The original concept of a sucrose ester-based detergent is due to Dr. Henry B. Hass who, in 1952, commissioned an investigation into the preparation of sucrose monoesters of fatty acids with Dr. Foster D. Snell (2). The impetus for this interest derived from a search for new, non-food uses for sucrose, sponsored by the Sugar Research Foundation, which was subsequently supplemented by a need for new markets for inedible animal fat, principally tallow (3). For example, the State Government of Nebraska commissioned research which led to the development of the Nebraska-Snell Process for the production of sucrose esters as an outlet for surplus tallow (4-5).

At about that time, concern was being expressed for the accumulation of detergent residues in inland water reserves, a consequence of the low rate of bacterial degradation of alkylarylsulphonates, then the principal surfactants used in industrial and domestic detergents. It had been shown that sucrose esters of cottonseed oil fatty acids were readily broken down by sewage bacteria (6), giving sucrose ester-based detergents a particular advantage. The switch to the more costly, linear alkyl sulphonates has, however, reduced the urgency of this problem in Europe (7), though the

demand for a totally biodegradable surfactant remains.
Sucrose esters of fatty acids are readily hydrolysed to sucrose and their component fatty acids by the
normal digestive enzymes, and show no evidence of toxicity. Consequently, they are of importance as food
additives (8), and in cosmetics (9-10). Many other
unique applications of sucrose esters have been recognised (11).

Sucrose stearates, palmitate, laurate and oleate
currently are manufactured by the Ryoto Company of
Japan (12) by the process originally developed under
the auspices of the Sugar Research Foundation (SRF).
Known as the 'Ryoto (formerly the Nitto) Sugar Ester
Process', it originally was put into operation by the
Dai Nippon Sugar Manufacturing Company Limited with a
capacity of 100 tons/month.   The S.R.F. process also
was licensed to companies in Germany, France, Italy
and Brazil.

The process originally selected involved transesterification between a triglyceride (fat, e.g., tallow)
and sucrose in dimethylformamide (DMF-a mutual solvent)
in the presence of a catalyst (potassium carbonate
$-K_2CO_3$) at a temperature of 90°C.   Under these conditions the fatty acid underwent transfer to sucrose
giving a complex mixture of mono- and diglycerides and
sucrose esters.   The reaction is not complete and the
product also contains unreacted sucrose and tallow,
giving problems in purification and analysis.

Methyl fatty acid esters, by-products of the production of glycerol from fats, are readily available.
Transesterification with sucrose, under conditions in
which methanol is removed continuously, results in the
equilibrium shifting towards complete reaction (13).

$$Ⓢ-(OH)_8 + MeO-COR \xrightarrow[\substack{DMF \\ 90°C}]{K_2CO_3} Ⓢ(OH)_7-O-COR + MeOH$$

Sucrose   Methyl   90°C   Sucrose ester
        fatty acid ester

Under these conditions, the formation of higher
esters of sucrose is favoured so that, even in the
presence of an excess of sucrose, the product will
contain 10% or more of the diester.   In practice this
rarely is a disadvantage.   Other solvents (dimethylsulphoxide, pyridine, N-acyl piperidine, etc.) and
catalysts can be used, but the separation and purification of the sucrose ester still is costly.

Owing to the relatively high cost of the purified
sucrose monoesters, they never were competitive with
petroleum based, anionic surfactants and their uses

are confined mainly to food and cosmetic applications.
For these applications, it is necessary to reduce the
level of residual, toxic dimethylformamide below around
50 ppm (14).
The avoidance of an aprotic solvent in the process
would offer a considerable economic advantage and several processes have been described using either a non-
polar solvent (xylene) (15), a non-toxic solvent (pro-
pane 1-2 diol) (4-5) or no solvent (16-17). With the
simultaneous ethoxylation of sucrose, a solvent is unnecessary, the molten fat and solid sugar undergoing
a heterogeneous reaction to give a complex,uniform mixture of products having detersive properties (18).
With the increasing cost and impending shortages
of petroleum-based products, the need for a surfactant
derived solely from regenerable raw materials is clear.
A low-cost route to sucrose fatty acid esters, the ad-
vantageous properties and applications of which are
now well recognised, would provide the necessary impetus. The Tate & Lyle (TAL) process (19) is designed
to provide this.

The TAL Process

In order to produce a sucrose-derived surfactant
which would be cost competitive with conventional
anionic and non-ionic surfactant-based detergents and
emulsifiers, it was decided to investigate the conditions under which sucrose would react directly with
a triglyceride in the absence of a solvent. This
would avoid the economic disadvantages of solvent loss
and recovery.
In this reaction it is well known that, unless
glycerol is continuously removed from the reaction
mixture, an equilibrium is reached in which chiefly
sucrose monoesters are present, together with mono-
and diglycerides, unreacted sucrose and triglycerides.
Potassium soaps also are present plus traces of glycerol, higher sucrose esters (mainly the di-) and catalyst. However, as the bulk of the remaining components possesses surface active properties, provided
that the di- and triglyceride content is kept to a
minimum, further separation of the product is not
necessary.
In practice it was found that, on heating a mixture of sucrose, tallow,and potassium carbonate for
several hours, the reaction product did exhibit detersive and emulsifying properties.
In order to optimise the conversion of sucrose to
its monoester without removing glycerol, which would

require operation under high vacuum, the factors
affecting the rate of reaction and the position of
equilibrium at ambient pressure were studied.   The
parameters considered were: 1. Temperature; 2. Ratio
of reactants; 3. Proportion of catalyst; 4. Nature of
catalyst; 5. Particle size of sucrose; and 6. Presence
of water.

Typically, the reactants and catalyst were heated
in an open vessel with mechanical stirring adequate
to give a uniform initial dispersion of reactants.
The reaction mixture was sampled at intervals for
analysis.

The composition of the mixture was determined by
quantitative gas liquid chromatography (glc) using
sucrose octaacetate as an internal standard.   A representative glc trace is shown in Figure 1.   From the
analysis it is possible to depict the course of the
reaction with time, in terms of the changes in the
relative concentrations of reactions and products.

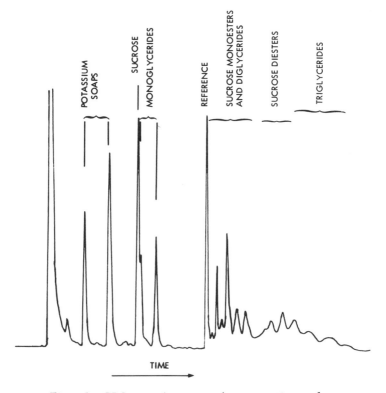

Figure 1.   GLC trace of sucrose surfactant reaction product

Temperature. The effect of temperature over the range 90 to 145°C was investigated. Above 130°C, discoloration and charring is unacceptably high, while below 115° C the rate of reaction is impracticably low. Between these two extremes, the optimal temperature is closely confined to 125 ± 5°C, as demonstrated in Figures 2 - 4.

Ratio of reactants. Approximately equimolar proportions of sucrose and tallow have been used in these studies. However, because the system is heterogeneous, the effect of changing the relative proportions of reactants will be lessened by their mutual insolubility. In practice, the highest conversion of sucrose to esters is obtained with an approximately 40% molar excess of sucrose.

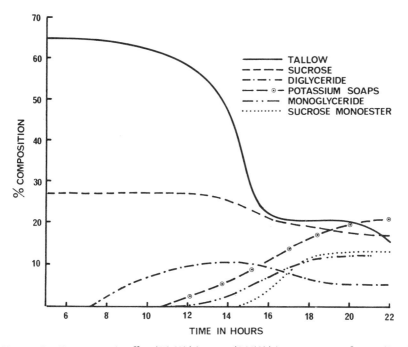

Figure 2. Reaction of tallow(27.4%)/sucrose(64.5%)/potassium carbonate(8.1%) at 115°C

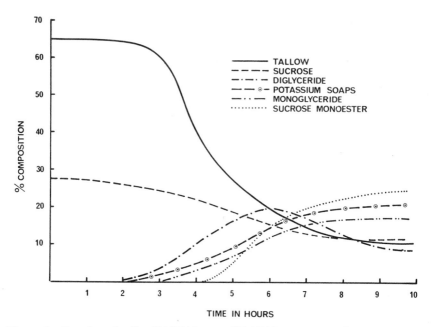

Figure 3. Reaction of tallow(64.5%)/sucrose(27.4%)/potassium carbonate(8.1%) at 125°C

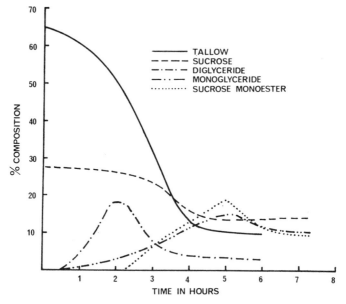

Figure 4. Reaction of tallow(64.5%)/sucrose(27.4%)/potassium carbonate(8.1%) at 135°C

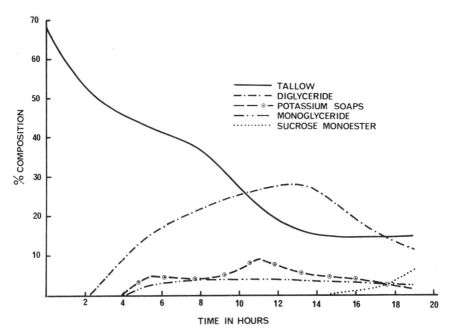

Figure 5. Reaction of tallow(68.7%)/sucrose(29.2%)/potassium carbonate(2.1%) at 125°C

Proportion of catalyst. The influence of the proportion of catalyst on the course and rate of the reaction was studied using potassium carbonate at levels of addition between 2.1% and 14.9% at 125°C. Again, since concentrations are indeterminable, the optimal addition of catalyst can only be established approximately. Nevertheless, from the results shown in Figures 4, 5, and 6, the low levels of catalyst result in negligible transesterification. At the highest levels of addition of catalyst, the equilibrium concentration of sucrose ester is proportionately increased but the viscosity of the mix becomes unduly high owing to the concomitant increased formation of potassium soaps. An 8.1% addition of catalyst gives an acceptable compromise.

Nature of catalyst. Of the active alkaline catalysts, which include the hydroxides of alkali and alkaline earth metals, alkali carbonates and bicarbonates, sodium methoxide, potassium phosphate and acetate, potassium carbonate proved to be the most effective catalyst.

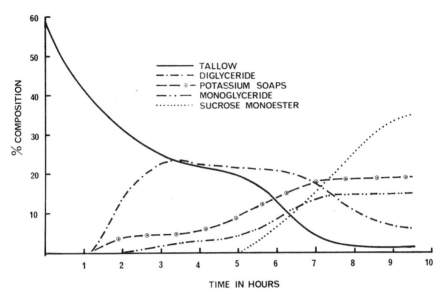

Figure 6. Reaction of tallow(59.7%)/sucrose(25.4%)/potassium carbonate(14.9%) at 125°C

Particle size of sucrose. The particle size of
both sucrose and catalyst is found to have little
effect on the rate of reaction.   Large granules are
difficult to suspend in the initial stages, while very
fine powders tend to agglomerate and remain undisper-
sed.   In practice, the sucrose and potassium carbonate
are milled to remove gross particles.

Presence of water. The presence of traces of water (< 1%) in the reaction mix is not found to have
any deleterious effect on the reaction and no special
precautions are necessary to exclude water.
As can be seen, at 125°C there is a 'dormant'
period of approximately 6 hours before any significant
reaction takes place.   This time has been termed the
initiation period, the end of which is associated with
the reaction mixture incorporating a large amount of
gas with concomitant foaming.   During this stage an
emulsion develops which is followed by the rapid formation of sucrose esters.   It is of interest to note
that the triglyceride reacts very slowly compared with
commercially available tallow.

The long initiation period has been the subject
of much detailed research within Tate & Lyle and it
appears that the first stage of the reaction is the
formation of a reactive intermediate, the diglycerides,
and it is these compounds which take part in the transesterification reaction with sucrose. A particular
advantage of the process is that the transesterification reaction virtually stops at the sucrose monoester
stage, resulting in only small amounts of diester being produced and no detectable amounts of unwanted,
higher esters.

The crude reaction product is now undergoing
extensive evaluation trials in a wide range of detergent applications with very encouraging results. A
measure of the confidence shown in the process is that
the material currently being evaluated is being produced on a pilot plant capable of producing up to
1000 tons/annum.

The TAL process is depicted in the flow diagram
(Figure 7). The raw materials are mixed under closely
controlled conditions in the presence of a catalyst.
The whole of the reaction mass is converted in a batch
reactor to a viscous, liquid surfactant which can be
cooled, solidified and flaked for incorporation in a
powdered detergent. Alternatively, it can remain in
the liquid state and be incorporated in a liquid detergent.

The major raw material for the process is a triglyceride (oil or fat). There is an added advantage
that many different kinds of triglycerides can be used
(Table I) which removed dependence on a single raw
material, which may fluctuate in price. Most of the
research and development, so far, has been based on
tallow which is readily available in the U.K. and which
traditionally is cheaper than imported vegetable oils.
Moreover, sufficient work has been done to ensure that,
in other parts of the world, locally produced triglycerides can be used in the process.

Table I

Triglyceride sources which can be used as feed
stock for sucrose surfactant production. There undoubtedly are many others which can be used.

Tallow	Soya Bean Oil
Palm Oil	Castor Oil
Coconut Oil	Rape Seed Oil
Cotton Seed Oil	Safflower Oil
Linseed Oil	Groundnut Oil
Palm Stearin Oil	

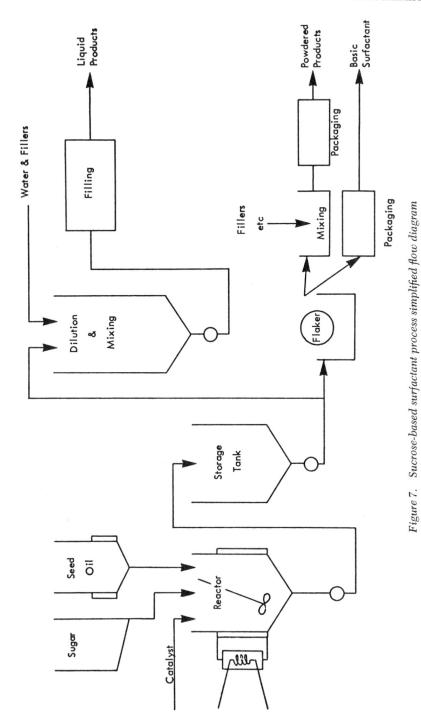

Figure 7. Sucrose-based surfactant process simplified flow diagram

Coincidental with the development of the new
low cost sucrose surfactant process, the upsurge in
world mineral oil prices has led to escalations in the
prices of petrochemicals and, consequently, the prices
of petroleum-based surfactants.   As an indication of
the increase, surfactant feedstock prices in Europe
have risen by 500% in the last four years.

Composition of TAL

The new Tate & Lyle sucrose surfactant is called
TAL, and is a complex mixture of sucrose esters, mono-,
di- and triglycerides and potassium soaps.

Table II

Typical Composition of TAL

Sucrose monoester	27	(%)
Sucrose higher ester	3	
Sucrose	13	
Triglycerides	3	
Diglycerides	9	
Monoglycerides	15	
Potassium soaps	30	

Its characteristics are unique and the mixture
cannot be used simply as a substitute for all other
surfactants.   The nearest petrochemical equivalent
surfactants are soft, non-ionic types.

Safety-in-Use

TAL has been tested extensively for safety in use.
TAL was found in feeding tests on rats to be completely
metabolized by the test animals, without any harmful
effects.   Because of this, it was not possible to
establish an LD_{50} but levels as high as 11.5 g/kg body
weight/day were fed to the animals without toxic
effects.
TAL is safe to use as it does not cause skin irritation nor cause allergenic reactions.   Skin irritation tests using standard laboratory methods have been
carried out on animals with no adverse results.
TAL is not toxic to plants and can be used safely
in agricultural formulations.   Phytotoxicity tests
have been carried out on peas and maize plants.   After
spraying with a 2% solution of TAL, the plants produced
flowers and fruits normally, as compared with the controls.

Compatibility

TAL is compatible with all conventional fillers, builders, dyes, perfumes and brighteners within a pH range of 7 to 12. Experimental formulations using non-phosphate builders are under evaluation and look promising.

Properties

The foaming properties of TAL are very low. This means that it can be formulated very simply into detergents for use in automatic machines (domestic and industrial) without the incorporation of foam suppressants.
An added advantage of such formulations is that TAL has inherent fabric softening properties which obviate the need for additional softeners and conditioners in clothes washing.

Table III

Some Properties of TAL

Appearance A tan, free flowing, flaked product which softens at 80°C and flows readily at 115° C.

Density 1.2g/ml at 25°C.
0.7g/ml at 125°C.

Solubility The maximum concentration obtainable in water is 43% w/w. This "solution" can be diluted with water readily.

Surface tension The standard lowering of surface tension of water is 32 dynes/cm at concentration of 200 ppm-300 ppm.

pH 9.7 in 1% solution.

Biodegradability 100%

Biodegradability of TAL

The complex nature of TAL has made it impossible to use standard procedures (20-21) for determining its biodegradability, for two reasons. Firstly, unreacted sucrose and tallow in TAL are both good substrates for microbial growth; hence observations that TAL will support microbial growth, in the absence of any other carbon source, are not indicative of its overall bio-degradability. Secondly, TAL cannot be determined by a single method; hence, analytical techniques have had to be developed for each component of TAL and pure standards synthesised.

We have used several, complementary approaches to measure TAL biodegradability based on the OECD screening test. These include the use of radioactively labelled detergent; measurements of the growth rates and maximum counts of microorganisms from natural sources on each component of TAL in a basal salts medium; and finally, glc analysis of each component in solutions incubated with soil suspensions. The evidence accumulated to date suggests that the components of TAL are readily metabolised by microorganisms. Work now is in progress to study TAL biodegradability in model, effluent disposal plants.

The surfactant breaks down rapidly and totally thus easing the pollution load on the environment. If TAL is discharged to a marine environment it has a low toxicity to marine life. Tests on the brown shrimp as an indicator organism give an LC_{50} value of 5000 ppm.

Evaluating the Surfactant Market

The U.K. market for surfactants is complex and highly structured. The investigation of this market on behalf of TAL has led to the development of a series of evaluating stages which can be applied to any market in which it is intended to sell sucrose sur-factants.

1. A review of the raw material supply situation for existing surfactants and for the sucrose-based products.

2. A comprehensive study of the local surfactant market.

3. An identification of the most significant market areas.

4. The formulation and testing of equivalent products to those in use in the chosen market sectors.

5. The selection and collaboration with suitable companies already using surfactants.

The underlying advantage of TAL is cost effectiveness. Secondary advantages become apparent according
to the market sector under consideration.

Biodegradability	–	Textile and wood industries.
Non-toxicity	–	Food, pharmaceuticals, oil slick dispersal.
Non-irritancy	–	Cosmetics, toiletries.
Non-allergenicity	–	Cosmetics, toiletries, domestic cleaners.
Low foaming power	–	Automatic machines.

Legal

The TAL manufacturing process is protected by
world-wide patents (19),   including a recently granted
US patent.   The name TAL is a trade mark belonging to
Tate & Lyle Limited.

Economic Assessment

When used for detergent formulations the competi-
tiveness of the surfactant depends essentially on the
relative costs compared with petrochemically based
products.   There are circumstances where one or more
of the other attributes of TAL could carry a premium,
e.g., biodegradability, fabric softening, non-toxicity,
but, essentially, the critical factor is price rela-
tivity.
Petroleum-based surfactants are derived from three
principal feedstocks, namely, ethylene, n-paraffins and
benzene.   Of these three feedstocks, the first has
had widest use since it is the lowest in cost (1973).
Benzene, on the other hand, generally has been higher
in cost than the other two feedstocks and has tended
to be used only when necessary.   It is interesting to
note that the relative cost differentials between the
three feedstocks are changing rapidly and, underlying
an increase in the absolute cost level of all petrochemical feedstocks, it is expected that the relative
costs of benzene and ethylene will be reversed by the
next decade.
The linear alkybenzene sulphonates (LAS), which
are already the "workhorse", active ingredients of the
industry, are derived from benzene and n-paraffins.
Prices have varied widely in the recent past due to
fluctuations in market conditions, governmental actions
in terms of price freezes and other factors.   In the

U.K. the price of LAS has varied between $630 and
$1,200 per ton.
The major cost factor for the sugar-based product
is the price of sugar and triglyceride.   In many
countries, for example the U.K., tallow is the most
suitable, low-cost source of triglyceride.   In other
countries, vegetable oils are more readily available
and may often be obtained at prices considerably below
the quoted, world market price.   It is of significant
advantage that the TAL process is sufficiently flexible
to utilise alternative triglyceride sources and it is
possible therefore to "play the market" in terms of
triglyceride input.
An examination of world sugar prices over the last
4 years might not suggest that this commodity has any
measure of price stability.   World prices moved from
a low of 5.58¢/lb in 1972 to a high of 56.63¢/lb during
1974.   More recently, the London daily price has returned to about $300/ton, but it is not expected to
return to the low levels which pertained up to 1972.
During a 20 year period prior to 1972, the world price
of sugar averaged $61 per ton.

Ester Separation

If the composition of the reaction mixture is considered in detail, it can be seen that it should be
possible to isolate different emulsifier systems from
this mixture by suitable separation procedures.   The
compositions of these new systems are shown below.

A) Sucroglycerides   B) Sucrose esters

    Triglycerides   Sucrose monoesters
    Diglycerides   Sucrose diesters
    Monoglycerides
    Sucrose monoesters
    Sucrose diesters

Again, it was considered that a separation system
was required that was relatively simple to operate
which used solvents that were already accepted as
additives in the food industry.   Though it could be
predicted that a liquid-liquid extraction procedure
would be extremely difficult to use, because of the
nature of the surfactant system, this method after investigation in detail, eventually was discarded in fav-
or of the preferred, solid-liquid extraction procedure.
The latter process is outlined in the schematic diagram in Figure 8.

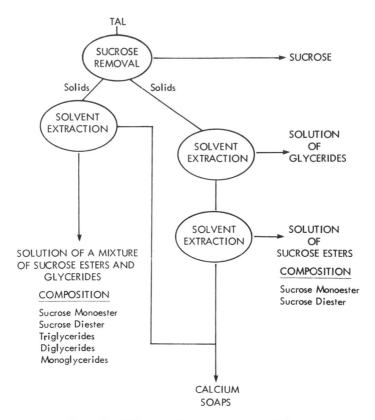

Figure 8.   TAL separation process for emulsifiers

    The all important step of this process is the pre-
cipitation of an insoluble soap.   This effectively results in precipitation of all the components of the
surfactant mixture with the exception of sucrose and
the water soluble degradation products.
    One of the major advantages of this total process
is that the physico-chemical properties such as H.L.B.
values, solubilities, wetting times, etc., of these
emulsifiers are governed largely by the nature of the
input material.   Therefore, by changing the nature
of the triglyceride used in the surfactant-producing
reaction, the properties of the sucroglycerides and
sucrose esters isolated from such a reaction mixture
can be altered considerably.   Consequently, a wide
spectrum of applications can be covered by essentially
one process, by altering the nature of the input material.

Abstract

Without the need of a sometimes costly and often toxic solvent employed in other processes, sucrose fatty acid esters can now be prepared by the direct reaction of sucrose and a triglyceride in the presence of potassium carbonate at 125°. Initially, the mixture is heterogeneous but towards the end of the reaction a single phase is noted, containing rougly equal amounts of sucrose esters and potassium soaps (60%), the remainder being unreacted sucrose and mixed glycerides.

Research has shown that the individual components of the mixture can be isolated singly or as required mixtures, by the techniques of solid-liquid extraction procedures. For example, sucrose esters are obtained after double decomposition of potassium soaps with a suitable metal salt, selective extraction of the mixed glycerides followed by extraction of the sucrose esters. These esters have great potential as emulsifiers, for example in the food and cosmetic industries, as they are completely non-toxic and biodegradable. Similarly, the very effective detergent action of a mixture of sucrose esters, soaps and sucrose (made simply by one step extraction of the crude reaction product), can be employed on a heavy tonnage scale in many detergent formulations without fear of polluting the environment.

Literature Cited

1. Berne-Allen, A. and Nobile, L.,(1965), Virginia J. Sci., 16, (3), 229-238.
2. Hass, H.B.,"Sugar Esters", Noyes Development Corp. 1968, 1.
3. Berne-Allen, A.,Fette, Seifen, Austrichmittel, (1965), 67, 509-511.
4. Osipow, L.I., U.S.P. 3,480,616.
5. Osipow, L.I. and Rosenblatt, W.,J. Amer. Oil Chem. Soc. (1967), 44, 307-309.
6. Isaac, P.C.G. and Jenkins, D.,Chem. & Ind. (London), (1958), 976.
7. Truesdale, G. A.,Chem. Prod., (1962), 25, 22.
8. Ames, G. R.,Tropical Science, (1962), 4, 64-73. Klis, J. B.,Food Proc. (1963), (July), 79-80.
9. Osipow, L.I., J. Soc. Cos. Chem., (1956), 7, (3), 249-255.
10. Nobile, L.,Drug and Cosmetic Ind. (1961), 89,(1).
11. Kollinitsch, V.,"Sucrose Chemicals", The International Sugar Research Foundation, Inc, Bethesda, (1970).

12. Japan Chemical Week, (1975), (September), 12.
13. Hass, H. B., Snell, F. D., York, W.C., and Osipow,
     L. I., U.S.P. 2,893,990.
14. "Codex Alimentarius", F.A.O./W.H.O., (1973), 38.
15. Curtis, G. W., U.S.P. 2,999,858.
16. B. P. 1,188,614; Jap.P. 14486/76.
17. Feuge, R. O., Zeringrie, Jr., H.J., Weiss, T. J.
     and Brown, M.,J. Amer.Oil Chem. Soc. (1970),
     47, 56-60.
18. B.P. 1,160,144; B.P. 1,099,777.
19. Parker, K.J., Khan, R.A. and Mufti, K.S.,B.P.
     1,399,053 and foreign equivalents.
20. Determination of the biodegradability of anionic
     synthetic surface active agents, Pollution by
     detergents, O.E.C.D., Paris, (1971).
21. Fifteenth Progress Report on the Standing Techni-
     cal Committee on Synthetic Detergents. Appen-
     dix G. " A method for assessing the biodegrad-
     ability of Nonionic Surfactants".   30-36.
     H.M.S.O., London, (1974).

Biographic Notes

Kenneth J. Parker, Ph. D., General Manager, Group
R & D. Educated at Oxford Univ. Joined Tate & Lyle
Ltd., 1955. Some 30 papers and patents on sugar, its
chemistry and technology. Tate & Lyle, Ltd., Philip
Lyle Memorial Research Laboratory, P.O. Box 68, Read-
ing, Berkshire RG6 2BA, England.

8

A Sugar Ester Process and Its Applications in Calf Feeding and Human Food Additives

LOUIS BOBICHON

Rhône-Poulenc S.A., 22 Avenue Montaigne, Paris, France

The process which we are using at Rhône-Poulenc
for the manufacture of sucroglycerides is not new.
It is the one which has been patented and licensed to
Société Melle-Bezons in 1958 by the Sugar Research
Foundation, Inc. This process was studied and developed, at that time in cooperation with Ledoga, and was
industrialized in 1963. Melle-Bezons has been acquired by Rhône-Poulenc and this is the reason why sucroglycerides actually are sold by the Fine Chemicals
Division of this company.
      Much already has been published on sucroglycerides
obtained by this process and a first review was made at
the International Symposium on Sugar Esters, held in
Paris in June, 1961 (1).
      A second review appeared in the papers presented
by Passedouet, Loiseau and Antoine (2-3) at the International Sugar Ester Symposium held in San Francisco,
in August, 1967.
      My task is to report the status of sucroglycerides
at Rhône-Poulenc, trying to cover all the aspects of
these products, including manufacture, composition,
specifications and applications.

Manufacture of Sucroglycerides

      To begin, for the manufacture of Rhône-Poulenc
(R.P.) sucroglycerides, we now are working a plant of
2000 T / year capacity. The process involves the reaction of sucrose with a triglyceride, such as tallow or
palm oil, and is performed in dimethylformamide with
potassium carbonate as a catalyst. This mixture is
heated and at the end of the reaction, the dimethylformamide is distilled off for recovery, and the reaction
product is purified by washing with an aqueous phase to
eliminate the last traces of dimethylformamide. The

final product is a viscous liquid when melted, and a
yellow waxy material when cold.

Composition of Sucroglycerides

The products of our procedure are mixtures of sucrose esters and glycerides, for which an average composition, obtained by using tallow or palm oil as a
starting material, is presented in Table I.

Table I.   Composition of R.P. Sucroglycerides

	Tallow	Palm Oil
Sucrose monoester	26	24
Sucrose diester	16	15
Monoglyceride	15	19
Diglyceride	23	25
Triglyceride	13	10
Free sugar	1	1
Free fatty acids	2	2
Combined fatty acids (soap)	3	3
Water	1	1

Composition has been determined by column chromatography.

Specifications

Specifications for our products may vary slightly
according to which glycerides are used in the process.
However, a good figure can be drawn from the specifications we certify for the quality sold in the human food
additive market.   These specifications are given in
Table II.

Table II.   Food Additive Standards of R.P. Sucrogly-
             cerides.

Acid value	not exceeding	6
Soap	"	6 %
Moisture	"	0.5 %
Free sugar	"	1.5 %
Saponification value		150 + 10
Sulphated ash	"	1 %
Dimethylformamide	"	5 ppm
Heavy metals	"	20 ppm
Arsenic	"	3 ppm

There is not much to be said, regarding the analytical procedures used to make these determinations, because these are well-known. May I mention, however, that: the percentage of soap is obtained by direct potentiometric titration of sucroglyceride dissolved in 2-propanol and water; the free sugar is determined by the thin layer chromatographic method in comparison with a standard; and the dimethylformamide content is obtained colorimetrically by the copper dimethyldithiocarbamate method.

Some work had to be done in the plant to meet these low specifications for soap and dimethylformamide, in order to comply with official specifications for these impurities. The main problem, in fact, has been to reduce the dimethylformamide content of our product, which was in the order of 100 ppm when we started the manufacture of sucroglyceride. We now can specify less than 5 ppm.

Sucroglycerides in Animal Feeding

The main use of our sucroglycerides, in the past and today, remains in the animal feeding market, where we sell our product under the trade name Celynol. We sell two products: Celynol MST 11, which is a tallow sucroglyceride; and Celynol TL, which is a mixture of tallow sucroglyceride and soya lecithin. Both of these have been approved in the animal feeding market by Belgium, England, France, Italy and Switzerland. They are used mainly in manufacturing reconstituted milk for calves. In this field they seem quite well suited because of their relatively cheap price and their desirable physical and biological properties. Their physical properties are easy to appreciate in the manufacture of milk replacers, especially in what we call the wet process, that is the process using the spray drying method. This method is used mostly in Europe. The sucroglycerides promote formation of a very good emulsion and decrease the viscosity of the mix before spray drying.

The sucroglycerides also have an anticaking effect as well as an antioxidant effect, thus helping to increase the storage time of the spray dried material. In addition they impart antifoaming properties to the dried milks and considerably reduce the formation of foam when the milks are reconstituted.

Their biological properties have been tested many times, and more recently by Robert (4). It was demonstrated that sucroglycerides decrease the consumption index, increase the digestibility of fats and nitrogen,

and decrease the incidence of diarrhea of half day
duration.   This is illustrated in Table III.

Table III.   Effect of Sucroglycerides on Calf Nutrition

	Without Emulsifier	With Sucroglycerides
Weight gain after 84 d (kg)	79.0	98.0
Daily weight gain after 84 days (gm)	941	1167
Consumption index	1.75	1.45
Fat digestibility index (5-6 week)	48.6	71
Nitrogen digestibility index (5-6 week)	88.7	91.8
Diarrhea 1/2 d after 84 d	14.4	2.7

Our sucroglycerides usually are sold in the melted
state, in tank cars, and are incorporated in the milks
in that melted state.   The most common method of incorporation is to add melted sucroglyceride to skimmed
milk kept at 60°C.   Once the sucroglyceride has been
dispersed, then the fat may be added and, after homo-
genization, the mixture may be spray dried.   The
dosage of sucroglyceride in this example is 1.5 to 2 %
of the fat used.

Sucroglycerides in Human Food Additives

In the human food additive market, we also sell
our sucroglycerides under the trade name Celynol and
our products have been approved in Belgium, England,
France, and Switzerland.   This is a market in which
we have tried to use some of the interesting characteristics of sucroglycerides, such as: the emulsifying
properties for both oil-in-water and water-in-oil emulsions,   with applications in margarine manufacture,
non-alcoholic beverages to incorporate aromatics, and
in sauces, and dressings; and the ability of sucroglycerides to obtain homogenous preparations in situations where the various components are difficult to
mix.   Here applications include:
- dispersion of coloring agents in paraffin or polymer
  materials for manufacture of cheese crusts,
- coatings of powders such as butter, vegetables, and
  fruits to facilitate subsequent mixing with other
  ingredients,
- manufacture of biscuits, pastry doughs or instant
  puddings to facilitate the mixing of ingredients such
  as flour, fat, sugar, etc.,

- the prevention of migration of ingredients in food confectioning such as ravioli and canneloni manufacture or chocolate confectionery,
- the reduction of viscosity in such instances as chocolate manufacture, where a lowering of manufacturing temperature has some effect on the stability of the finished product,
- the reduction in the size of the crystals formed in liquid or semi-liquid media during the course of solidification in the presence of a fatty or non-fatty phase, for instance in ice cream manufacture and confectionery.
- the improvement of texture, particularly in baked products,
- the ability to restore the whipping properties of such products as egg powder, once the eggs have been dried.

Conclusion

The sucroglycerides are relatively cheap materials, they are non-toxic and exhibit interesting emulsifying properties. They have been approved by the European Committee as additives for application in the food industry under the code number E 474 (5). They definitely have a potential in the food and feed industries.

Abstract

Rhône-Poulenc manufactures sucroglycerides by the Sugar Research Foundation process which involves the use of dimethylformamide as a solvent, which means that the product must be purified in order to meet the low requirements for dimethylformamide. This product has a high content in mono- and diester of sucrose, a low content in sucrose and low in soap. It is used mainly in the animal food field in reconstituted milk for calf feeding because of its tensioactive properties. Other applications in human markets have been found.

Literature Cited

1 - International Symposium on Sugar Esters, Maison de la Chimie, Paris, 1961.
2 - Passedouet, H., Loiseau, B., and Antoine R., "New prospects in animal feeding:Sucroglycerides", International Symposium on Sugar Esters, San Francisco, 1967.
3 - Passedouet, H., Loiseau, B., and Antoine, R., "Applications of sucroglycerides in food",

International Symposium on Sugar Esters, San
Francisco, 1967.
4 - Robert, J., "Influence de l'adjonction de sucroglycerides á un aliment d'allaitement sur les performances zootechniques du veau de boucherie et la
digestibilité des différents élements de la ration"
Les Industries de l'Alimentation animale, 1973,
No. 7-8.
5 - "Projet de Directive de la CEE", J.O. CEE, (1969)
No. 5-54/1, April, 28.

Biographic Notes

Louis Bobichon, Dir. Biochemical and Fermentation
Sciences.   Educated at Ecole de Chimie Industrielle
de Lyon.   Joined Rhône-Poulenc in 1945.   Rhone-Pou-
lenc, Research and Development, 22 Avenue Montaigne, ·
F75360, Paris, France.

Sucrose Esters in Bakery Foods

P. A. SEIB, W. J. HOOVER, and C. C. TSEN

Department of Grain Science, Kansas State University, Manhattan, Kans. 66506

Surfactants are used in foods to accomplish one or
more of the functions (1) given in Table I.   Most of
the functions are encountered in the bakery.   For example, surfactants are added to cake premixes to prevent solid ingredients from clumping during hydration.
Also, most cake formulas include emulsifiers to improve
batter aeration and the eating and keeping qualities of
cake.   Surfactants complex with wheat proteins in
bread dough to strengthen dough, while in bread crumb
they apparently modify crystallization properties of
starch, and thereby prevent bread from firming.

Table I.   Functions of Surfactants in Foods.

Wetting	Complexing
Emulsifying	Modifying Crystals
Foaming	Suspending

A wide choice of surfactants (Table II) is available to the food technologist (1, 2).   The choice depends on many factors but the most important are listed in Table III.   Except for food-safety, the factors
are self-explanatory.   In the United States, the Food
and Drug Administration (FDA) permits white bread and
white enriched bread (3), for example, to contain a
maximum of 0.5 part by weight of dough conditioner per
100 parts of flour.   The baker may use a dough conditioner from the surfactants listed either in the
"Generally Recognized As Safe" (GRAS) section (4) or in
Subpart D (5) of the Food-Additive Regulations.   In
either case, he must follow "good-manufacturing-prac-
tices" (GMP), which is to say, the dough conditioner
should be added in an amount "...reasonably required
to accomplish its intended physical, nutritional, or
other technical effect...".   Surfactants on the GRAS

list (and some of those in Subpart D), such as diacetyl
tartaric acid esters of monoglycerides of edible oils
(fats) may be added to any food under GMP, others in
Subpart D are permitted only in certain foods at speci-
fied maximum levels (for example, polyoxyethylene sorbitan monostearate).

Table II.   Surfactants Used in Foods in the
            United States of America

    Lecithin
    Glycerol and Polyglycerol Esters
    Propylene Glycol Esters
    Lactylate Esters
    Sorbitan Esters
    Fumarate, Tartrate, Succinate Esters
    Ethoxylated Derivatives

Table III. Factors Affecting Choice of Food-grade
            Surfactant.

    Unique Function   Handling; Storage
    Two or more Functions   Sanitation
    Cost   Safety

    Sucrose esters have not been approved for use in
the U.S.A.   However, they were included on the list of
food emulsifiers compiled by the Codex Alimentarius
Committee of the Food and Agriculture Organization/
World Health Organization.   In addition, a 1974 directive drafted by the nine countries of the European Economic Community (2) placed sucrose esters with Annex
I substances.   Annex I substances are those likely to
be approved for food use by all states of the EEC.
To the authors' knowledge, sucrose esters are currently
used in foods in Japan, Switzerland, France, Belgium
and England.

Surfactants in Bakery Foods

    As previously intimated, food-grade surfactants
are important in the production of bakery foods.
Table IV shows the annual production of bakery foods
in the United States.   Yeast-leavened, bread-type products (bread, rolls, and sweet-goods) dominate the market, accounting for ∿ 70% of the total poundage.   White
pan bread is the single most important bakery food
(Table V), and each one-pound loaf contains up to 1.4g
added surfactant, or a total potential market in breadmaking of 90,000 lb day^{-1} (4.1 MT/day).   Use of dough

conditioners in breadmaking has been estimated (6) to lower the retail price of bread 10-15%. Because of the prominent position of bread, our discussion of food-grade surfactants in the bakery deals heavily with bread.

Table IV. Bakery Foods in the United States of America Annual Production, Billions of lb.

Bread	13.4	Sweet Goods	1.3
Rolls	4.0	Pie	0.9
Cookies & Crackers	4.0	Donuts (Cake)	0.7
Cakes	1.8	Donuts	0.4

Table V.   Most Important Bakery Food

White Pan Bread = 50% of Bakery Foods

Surfactant, 1.4 g in lb of bread = 90,000 lb d^{-1}

Surfactants are used in breadmaking for three major reasons (Table VI).

Table VI.   Functions of Surfactants in Breadmaking.

Dough Strengthener	Gas Cells
Oven Spring	Non-Wheat Protein
Crumb Softening	Replace Fat

Depending on chemical structure, a surfactant will: (a) prevent dough from collapsing during processing (strengthen dough); (b) increase oven-spring of a loaf during initial stages of baking, thereby increasing loaf-volume while simultaneously keeping gas-cells of the bread crumb small; and (c) prolong shelf-life of bread by softening the crumb. As an example of how chemical structure affects the functions of a surfactant, sodium stearoyl 2-lactylate (SSL) performs all three functions (7), but glycerol monostearate (GMS) and polyoxyethylenesorbitan monostearate (20) perform functions (c) and (a), respectively (8). Surfactants with dough-strengthening properties also produce doughs that hold more water and have less sticky surfaces. Strong, dry-surfaced doughs mean many fewer "crippled" loaves,while the extra water generates more product (white bread contains < 38% water). Dough-strengtheners also can be used to raise protein and/or lower fat in loaves (9-10). Before the volatile commodity prices of recent years, bread dough normally contained 3% lard or short-

ening,(percentages are given as "baker's percentages",
i.e. 1% fat in dough means one part by weight of fat
has been added per 100 parts of flour).   Shortening
improves loaf volume, slicing, and shelf-life of bread.
But, dough strengtheners at < 0.5% together with vegetable oils at 0-1%, can be used to replace shortening.
Such ingredient interchanges help reduce costs.
        High protein breads are of interest in developing
countries to alleviate protein/calorie malnutrition
(9-10). Bread provides an ideal vehicle to improve
nutrition because it is used widely and it can be fortified at the mill or bakery.   Adding protein-rich
flours from pulses and legumes to wheat flour results
in bread with low volume and poor texture.   Fortunately, dough strengtheners overcome those obstacles.
        Wheat flour contains 0.2% free polar lipid,approx-
imately one-half of which is digalactosyl diglycerides
(11).   Despite their low concentration, free-polar
lipids are essential to the breadmaking characteristics
of wheat flour (12).   Monoesters of sucrose contain
the same ratio of hexose to fatty acid as found in
digalactosyl diglycerides (Figure 1).

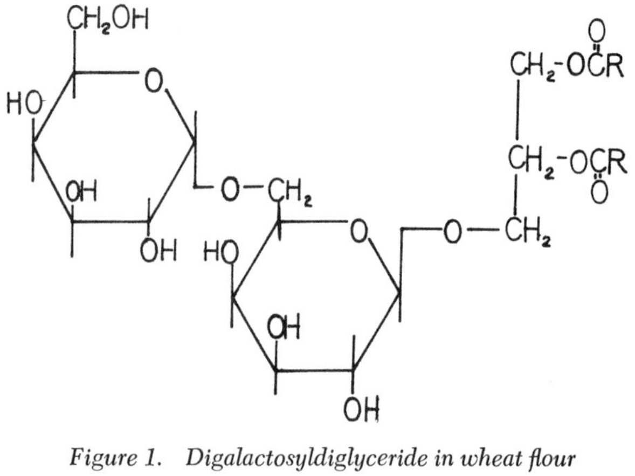

Figure 1.   Digalactosyldiglyceride in wheat flour

        The ability of sucrose esters to carry "foreign"
protein in bread is illustrated in Figure 2.   The figure shows pup loaves baked in an experimental breadmaking system developed during the past 30 years by
Professor Karl Finney of the Hard Winter Wheat Quality
Laboratory of the United States Department of Agriculture.   In this bake test (13), the baker is looking
for the highest possible loaf-volume accompanied by

Figure 2. Added lipids in breadmaking. Center pup-loaf, no added lipid; right, 0.75% commercial sucrose monopalmitate; left, 3% shortening (percent based on flour in formula). (Courtesy K. F. Finney, Grain Marketing Research Lab., US Dept. of Agriculture, Manhattan, Kansas)

the smallest and "best developed" gas cells. Loaf volumes in the bake test are reproducible to ± 2% when triplicate loaves are baked.

It is necessary to digress here to point out that loaf-volume potential of a bread formula is to the baker what acid strength of a compound is to the chemist. Loaf-volume can be decreased easily just as acidity can be lowered by dilution of a strong acid, but poor loaf volume produced by a formula under optimum conditions cannot be improved, just as acetic acid can never give a strongly acidic solution in water.

Returning to Figure 2, all three loaves were prepared using a composite flour containing ∿ 11% defatted, toasted soy flour. The loaf in the center, containing no added lipid, was unacceptable. The loaf on the left, containing 3% shortening (bakers' percentage) was improved over the control loaf, while the loaf on the right, containing ∿ 3/4% commercial sucrose monopalmitate, was highly acceptable. Loaves in Figure 2 illustrate the ability of fatty acid esters of sucrose to (a) carry soy flour, (b) spare shortening, and (c) strengthen dough (improved gas retention with faster proofing).

The sucrose ester used in Figure 2 was a commercial sample containing a mixture of compounds. The
manufacturer reported the ester consisted of 70% mono-
and 30% higher esters of a mixture (7/3, w/w) of pal-
mitic and stearic acids. The question arises as to
which chemical constituent(s) is (are) responsible
for the improving action of sucrose esters in breadmaking. Such information is needed to formulate the
most effective sucrose ester.

Sucrose Esters of Pure Fatty Acids

    In our work we used the Lemieux modification (14)
of the Hass procedure to produce sucrose esters of
pure fatty acids (capryllic to arachidic, including
oleic acid). We used silica gel chromatography to
isolate and purify the monoester and, in some cases,
the diester fractions. The pure esters were characterized by saponification number and sucrose content.
They had no effect on the production of CO_2 in a yeast-
flour ferment.
    The data in Table VII show that sucrose monopalmi-
tate (G14) was very active in overcoming the volume-
depressing effect of soy flour in bread, whereas the
diester fraction behaved passively. The control
loaf's (0% lipid) volume was 713 cc, which increased
dramatically to 851 cc with only 0.25% sucrose monopalmitate. The volume of the loaf made from 3% shortening was 880 cc, which could be readily surpassed
using 0.5-0.75% sucrose ester. The dipalmitate fraction had no volume-improving or detrimental effect at
0.25%. We concluded that only the monoesters of sucrose are dough conditioners in breadmaking.
    We also examined the effect of fatty acid chain
length on the dough-conditioning effect of sucrose
monoesters. Again, we used the volume-improving
action of sucrose monoesters in soy-fortified breads to
measure the dough-strengthening effect. Results are
shown in Figure 3.
    In this experiment, the volume of the control loaf
containing 3% shortening and no sucrose ester was 880cc
compared with 638 cc for the loaf with neither shortening nor sucrose ester. Adding 0.2 - 0.4% of a commercial sucrose ester (F-160, Dai-Ichi Kogyo SeiyaKu
Co., Ltd., Kyoto, Japan) dramatically improved loaf
volume, but additional amounts (0.4 - 0.8%) produced
no further improvement.
    Up to now (August, 1976) we have obtained only
three data points on the volume improving effect of the
pure sucrose monoesters. But those data points, along

Table VII. Effect of Lipids on Loaf-volume of Bread (cc) Prepared from 100 g of a Mixture (9:1) of Wheat Flour and Toasted Defatted Soy Flour.[a]

Lipid[b], %	Lard	Monopalm.[c]	Dipalm.
0	713	713	713
0.1	–	815	698
0.25	–	851[d]	688
3.0	880[d]	–	–

(a) In collaboration with H. Chung, K. Finney, and C. Magoffin, U. S. Grain Marketing Research Lab., Manhattan, KS.

(b) Percent based on weight of flour (14% H_2O).

(c) Sucrose monopalmitate.

(d) Crumb structure judged satisfactory.

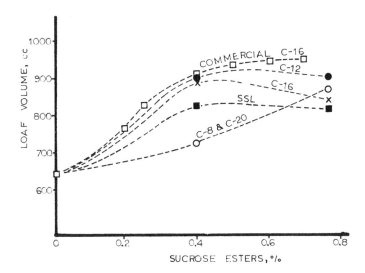

Figure 3. Performances of a commercial sucrose monopalmitate and of sucrose monoesters of pure fatty acids in bread (pup-loaf) made from a no-shortening formula containing a 9:1 mixture of wheat-soy flour. The loaf with 3% shortening had a loaf-volume of 880 cc. (Results determined by C. Magoffin and K. F. Finney, Grain Marketing Research Laboratory, U.S.D.A., Manhattan, Kansas, U.S.A., in collaboration with H. Chang, Dept. of Grain Science, Kansas State University, Manhattan, Kansas)

with the general shape of the curve for the commercial
ester, can be used to suggest shapes of curves for
the pure esters.   Figure 3 shows that at 0.4%, sucrose
monolaurate (C-12) and monopalmitate were much more
effective dough strengtheners than the monocapryllate
(C-8) or monoarachidate, (C-20).   Although not shown,
the monomyristate, monostearate, and monooleate esters
were almost as effective as the monolaurate ester, but
the monocaprate ester was somewhat less effective.
     Increasing the weight percentage of pure sucrose
ester in the pup-loaf formula from 0.4% to 0.75% produced some surprises.   At those levels, sucrose monocapryllate, monocaprate, and monoarachidate became more
effective than sucrose monomyristate, monopalmitate,
monostearate, and monooleate.   Only sucrose monolaurate did not decline in effectiveness; it showed a
broad, optimum volume-response with concentration in
dough.
     Two other points are evident in Figure 3.   Sodium
stearoyl 2-lactylate (SSL) is inferior to sucrose
esters for carrying soy protein in a no-shortening
bread formula.   Also, the superior performance of the
commercial sucrose ester implies   synergistic effects
in which a mixture of sucrose esters is somewhat more
effective as a dough strengthener than is a purified
monoester.

Carbon-13 Nuclear Magnetic Resonance Spectra

     A monoester of sucrose prepared from a pure fatty
acid methyl ester by transesterification is a mixture
of positional isomers (15-20).   Each isomer could be-
have differently in bread.   Data in Figure 4 show a
dramatic example of how positional isomers function
differently in breadmaking (21).   L-Ascorbyl 6-palmi-
tate is an excellent dough strengthener, which showed
a +92 cc volume response above a no-shortening control
loaf (905 cc).   On the other hand, L-ascorbyl 2-palmi-
tate decreased loaf volume 165 cc below the control.
     To tailor-make sucrose esters for food, it is im-
portant to learn which positional isomers are most effective in a given application. To begin such a program,
a rapid method of characterizing a mixture of isomers
of sucrose monoesters was needed. Preliminary results
indicated carbon-13 nuclear magnetic resonance (^{13}C
n.m.r.) can provide the required compositional information.
     A partial spectrum of the monoester fraction of
sucrose palmitate in methyl sulfoxide is shown in Figure 5.   The three signals of the anomeric carbon of

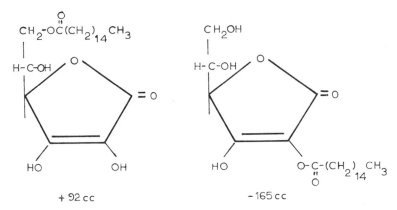

Figure 4. L-Ascorbyl 6- and 2-palmitate

the fructose residue indicated that the fraction contained three predominant isomers. That conclusion was confirmed by the appearance of three relatively intense carbonyl resonances at approximately 173 ppm. As primary hydroxyls normally are more readily esterified than secondary hydroxyls, it is assumed that the three main isomers are sucrose 6-palmitate, 6'-palmitate, and 1'-palmitate (primed positions are on the fructose residue).

The chemical shifts of the twelve carbons in sucrose have been assigned by three groups of investigators (22-24). Furthermore, Horton et al., found that the chemical shifts of sucrose in water were practically identical to those in methyl sulfoxide (23). We therefore, can compare the data on sucrose esters in methyl sulfoxide to those found in water.

The F-2 signal of lowest intensity (101.7 ppm) seen in Figure 5 was shifted approximately 2 ppm upfield from the two other F-2 signals (103.5 and 103.8 ppm). The chemical shift of a carbon in the β-position to a carbinol undergoes approximately a +2 ppm shift when the carbinol hydroxyl is esterified (25). The peak at 101.7 ppm thus can be assigned to the F-2 carbon of sucrose 1'-palmitate.

Two signals were observed in the F-4 region of the spectrum of sucrose monopalmitate. The F-4 carbon of sucrose is easily identified in the spectrum of sucrose, as it is the signal farthest downfield except for the anomeric carbons. The predominate signal at 82.4 ppm is close to the F-4 signal of sucrose (82.2 ppm). Therefore, we assume that signal results from the overlapping resonances of the F-4 carbons of sucrose 6-palmitate and 1'-palmitate. The signal at 79.3 ppm

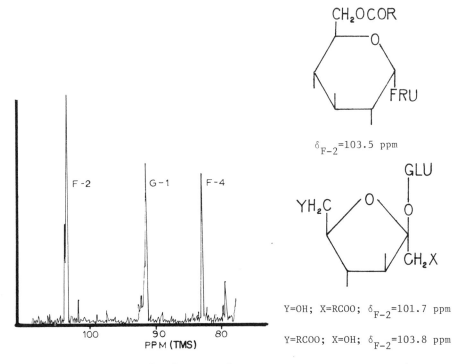

Figure 5. *Partial carbon-13 nuclear magnetic resonance spectrum of sucrose monopalmitate in methyl sulfoxide; pulse time was 3 sec with a total of 32,000 pulses*

was assigned to F-4 of sucrose 6'-palmitate. Reist and coworkers (26) previously reported the F-4 carbon in sucrose 6,6'-di-p-toluenesulfonate resonated at 80.8 or 79.7 ppm in acetone-d_6 solution, which is 1.4 - 2.5 ppm upfield from the position of the F-4 carbon of sucrose in water. Excluding solvent effects, esterification at the 6-position of fructose appears to cause an upfield shift of the F-4 signal.

The relative intensities of the F-2 and F-4 signals were used to calculate that the monoester fraction contained 64% sucrose 6-palmitate, 29% sucrose 6'-palmitate, and 7% 1'-palmitate, which agrees with conclusions of others who have reported that the predominate isomer is the 6-ester (17-20) and that the 1'-position is the most sterically hindered (26).

Crumb Softening

Crumb softeners are used by bakers to extend the shelf-life of bread, and thereby reduce delivery costs.

The crumb-softening power of a surfactant often is determined by measuring the weight required to compress bread crumb a given distance. Shortening (3% based on flour) gives some crumb softening (Figure 6), whereas sucrose monostearate (0.75% based on flour) softens bread crumb almost as well as sodium stearoyl 2-lactylate. The latter surfactant is as effective as the α-glycerol monoesters of fatty acids, the most widely used, commercial bread softener.

The softening power of sucrose monoesters declined progressively as the chain length of the fatty acid was decreased from C-18 to C-8 at both the 0.75% and 0.4% level. When plotted, the data produces a family of curves (Figure 6), all of which are not shown. Also, sucrose monooleate, monoarachidate, and sucrose monostearate were equally effective at 0.75%, but the first two were inferior at 0.4%.

Other Potential Bakery and Cereal Uses of Sucrose Esters

Soft cakes generally are prepared from formulas containing 40 - 70 parts of shortening per 100 parts of flour. Most shortenings are the emulsified type containing 3-6% of a surfactant,such as mono- and diglycerides, propylene glycol esters, or polyoxyethylene sorbitan monostearate. Surfactants function in cake to improve cake volume, produce a finer grain, carry

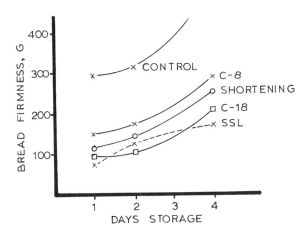

Figure 6. Crumb-softening effects of sucrose monoesters of pure fatty acids in soy-fortified bread (see Figure 3). Control contained no added lipid.

more sugar (high ratio cakes), and to improve eating
and keeping qualities of cake.   The total potential
market for surfactants in cake is approximately 10,000
lb day $^{-1}$.

Sucrose esters now are used in sponge cake in
Japan (27).   However, no information is available on
how they would perform in the variety of cakes made in
the U.S.A.

A particularly exciting market for sucrose esters
and other food surfactants could be developed if they
reduced the requirement for shortening in cakes, cookies, and crackers.   That would permit those popular
foods to be enjoyed with less caloric intake.   Tsen
(28) has reported, for example, that 25% of the shortening in cookies can be replaced if one uses < 1% surfactant.

Sucrose esters could be blended with protein-for-
tified flour for purchase by volunteer agencies overseas under Title II of the PL-480 program of the United
States Department of Agriculture.   In 1975, 150 million pounds of soy-fortified flour was shipped from
the United States to developing countries.   Currently
sodium stearoyl 2-lactylate is the approved surfactant
in soy-fortified flour.   Developing countries may
present still another market for sucrose esters.
Frequently those countries want to reduce imports of
wheat flour for bread by compositing wheat flour with
indigenous starchy flours from corn, sorghum, rice,
or cassava.   Surfactants are required to produce
acceptable breads from such composite flours (29).

The possibility of using sucrose esters in the
following applications should be explored; (1) in
hamburger and hot dog rolls to reduce the gluten added
(30); (2) to make pizza dough easier to sheet; (3)
in whipped toppings and fillings for cakes and sweet
goods; (4) to substitute for hardened fats in contin-
uous-mix bread to increase strength of the loaf's side-
walls; and (5) in pasta and noodles as well as precook-
ed breakfast cereals to reduce stickiness and sogginess
upon hydration.

Summary

Sucrose monoesters of common fatty acids are excellent dough conditioners in breadmaking.   They produce drier, stronger doughs; permit production of highly acceptable bread from wheat flour containing non-
wheat, high-protein, or high-starch flours; and they
spare shortening.   In addition, sucrose esters rank
closely to the most effective crumb softeners currently

used.   Although bread provides by far the largest
potential market for sucrose esters in the bakery,
other uses would be significant, including: emulsifier
for soft cakes; shortening-sparing agent in cookies and
crackers; gluten-sparing agent in buns; and emulsifier
in toppings and fillings.

Acknowledgments

The authors thank the International Sugar Research
Foundation, Inc. for financial support, and Dr. Richard
Loeppky of the University of Missouri, Columbia, for
the carbon-13 NMR measurements.

Abstract

In 1969, Pomeranz, Shogren and Finney, Cereal
Chemistry, 46, (1969), 503 and 513, reported protein-
fortified breads can be prepared using selected fatty-
acid esters of sucrose.   Additional useful functions
of sucrose esters in baked foods have been demonstrated
including, among others, shortening-sparing in bread,
cookies and crackers.   Additional studies are under-
way to determine the dough conditioning and crumb softening effects of sucrose esters in breadmaking.   A
review of the potential bakery market for sucrose
esters is presented.

References

1.   Griffin, W.C. and Lynch, M.J. in "Handbook of Food
      Additives," Chapter 10, p. 413, Chemical Rubber
      Company, Cleveland, Ohio (1968).
2.   Lauridsen, J.B., J. Am. Oil Chemists' Soc., (1976)
      53 400.
3.   Code of Federal Regulations, Title 21, Food and
      Drugs, Chapter I, Subchapter B, Part 17-Bakery
      Products, pp. 44-8, April 1, 1976.
4.   Ibid., Subpart B, Paragraph 121.101, p. 313.
5.   Ibid., Subpart D, Paragraphs 121.1000 through
      121.1267, pp. 342-420.
6.   Angeline, J.F., and Leonardos, G.P., Food Techno-
      logy, (1973), 27, (4), 40.
7.   Hoseney, R.C., Hsu, K.H., and Ling, R.S., Bakers
      Digest, (1976), 50, (4), 28.
8.   Langhans, R.K., and Thalheimer, W.G., Cereal Chem.,
      (1969), 46, 503 and 512.

9.   Pomeranz, Y., Shogren, M.D., and Finney, K.F.,
     Cereal Chem., (1969), 46, 503.
10. Tsen, C.C., and Hoover, W.J., Cereal Chem.,(1973),
     50, 7.
11. Mason, L.H., and Johnston, A.E., Cereal Chem.,
     (1958), 35, 435.
12. Hoseney, R.C., Finney, K.F., Pomeranze, Y., and
     Shogren, M.D., Cereal Chem., (1969) 46, 606.
13. Finney, K.F., Cereal Chem., (1945), 22, 149.
14. Lemieux, R.U. and McInnes, A.G., Can. J. Chem.,
     (1962), 40, 2376.
15. Weiss, T. J., Brown, M., Zeringue, H.J.,Jr., Feuge,
     R.O., J. Am. Oil Chemists' Soc., (1971), 48,
     145.
16. Lemieux, R.U., and McInnes, A.G., Can. J. Chem.,
     (1962), 40, 2394.
17. York, W.C., Finchler, A., Osipow, L. and Snell,
     F.D., J. Am. Oil Chemists' Soc., (1956), 33,
18. Gee, M., and Walker, H.G, Jr., Chem. and Ind.
     (1961), 829.
19. Reinefeld, I., and Klandianos, S., Zucher, (1968),
     21, (12) 330;
20. Reinefeld, I., and Klandianos, S., Chem. Abst.
     (1968), 69, 106989u.
21. Hoseney, R.C., Seib, P.A., and Deyoe, C.W., Cereal
     Chem., In Press.
22. Allerhand, A., and Doddrell, D., J. Am. Chem. Soc.
     (1971), 93, 2777.
23. Horton, D., Binkley, N.N., and Bhacca, N.S., Carb.
     Res., (1972), 23, 301.
24. Hough, L., Phadnis, S.P., Tarelli, E., and Price,
     R., Car. Res., (1976), 47, 151.
25. Levy, G.C., and Nelson, G.L., "Carbon-13 Nuclear
     Magnetic Resonance for Organic Chemists,"
     Wiley-Interscience, p.47, New York (1972).
26. Almquist, R.G., and Reist, E.J., Carb. Res.,(1976)
     46, 33.
27. Kosaka, T. "Sucrochemistry Symposium", ACS Sympo-
     sium Series, Chapter 6, Washington, D.C., 1977.
28. Tsen, C.C., Peters, E.M., Schaffer, T.,and Hoover,
     W.J., Bakers Digest, (1973), 47, (4), 36.
29. Tsen, C.C., Medina, G., and Huang, D.S., "Proceed-
     ings of the Fourth International Congress of
     Food Science and Technology", Institute of
     Agronomy and Grain Technology, Valencia, Spain,
     In Press.
30. Jackel, S.S., Diachuck, V.R., and Neu, G.D.,
     Bakers Digest, (1974), 48, (1), 40.

Biographic Notes

Professor Paul A. Seib, Ph.D., Dept. of Grain Science and Industry. Educated at Purdue Univ. Joined staff of the Inst. of Paper Chem. 1965-70 and since at Kansas State Univ. Some 20 papers on carbohydrate chemistry. Department of Grain Science and Industry, Kansas State Univ., Manhattan, Kansas 66502 U.S.A.

10

Method for Preparing Esters of Polyalcohols (Sugar Alcohols)

J. A. VAN VELTHUIJSEN, J. G. HEESEN., and P. K. KUIPERS

C.V. Chemie Combinatie, Amsterdam C.C.A., Arkelsedijk 46, Gorinchem, Holland

Esters of polyols (polyalcohols) and higher fatty acids frequently are used as non-ionic emulsifying agents. The sorbitan esters of fatty acids are well known. They are prepared by esterification of sorbitol with fatty acids at temperatures of at least 190°C, usually at 220-240°C. During the reaction, a substantial portion of the sorbitol (more than 75%)is converted into anhydro-compounds (intramolecular ethers like sorbitan and isosorbide). The products become less hydrophilic due to the formation of these anhydro-compounds.

Polyol esters of fatty acids can be prepared without simultaneous formation of anhydro-compounds by reesterification of the methyl or glyceryl esters in a polar solvent, in the same way that sucrose esters have been prepared. Residues of the toxic solvents are difficult to eliminate from the final products,and this creates problems for the products to be used in foods.

Now we have found that polyol esters, containing a minimal percentage of anhydro-compounds, can be obtained if the fatty acid is esterified with the polyol or a glycoside thereof, in the presence of a fatty acid soap (in a quantity of 10% or more calculated upon the polyol) at a temperature between 100°C and 190°C with simultaneous elimination of the water formed during the reaction. If, for instance, sorbitol is esterified at 150°C in the presence of fatty acid soap, only 3 to 5% of the sorbitol is converted to anhydro-compounds (sorbitan). (See Figure 1.)

It is a well-known procedure to use soap as a catalyst or as a miscibility promoting agent in preparing sucrose esters of fatty acids without a solvent. One of these is the microemulsion process for preparing sucrose esters,developed by Snell for the State of Nebraska. Soap also is deemed an essential part of

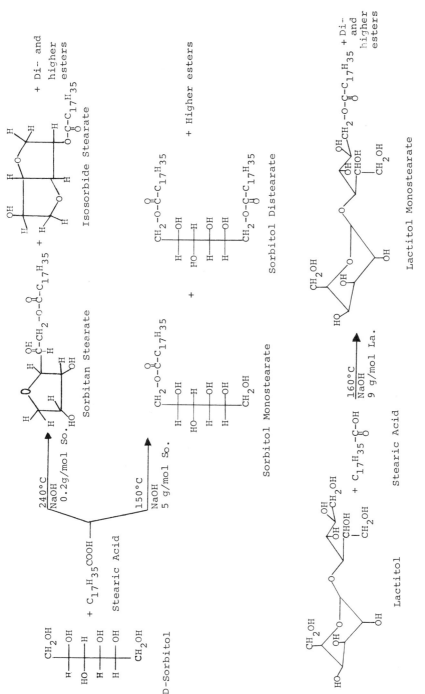

Figure 1

the reaction mixture in the molten sucrose process developed by Dr. R.O. Feuge of Southern Regional Research
Laboratory of the U.S.D.A.   In our method for the
preparation of polyol esters, a good result generally
can be obtained with less soap, and no homogeneous
solution or microemulsion is required for the esterification of the greater part of the fatty acid.
    In our process, the polyols may be added to the
reaction mixture as an aqueous solution.   However,
water is not an essential component of the mixture as
it is in the microemulsion process for the preparation
of sucrose esters of fatty acids.
    A characteristic difference from the preparation
of sucrose esters, is that sugar alcohols are much
more stable than sucrose, so that they can withstand
the relatively long reaction times needed for direct
esterification by fatty acids.   The method is very
satisfactory for preparing fatty acid esters of sugar
alcohols of mono- and disaccharides.
    Sugar alcohols can be prepared by hydrogenation
of reducing mono-, di- and oligosaccharides.   Sorbitol
and mannitol are obtained from sucrose, maltitol from
maltose and lactitol from lactose.   Maltose is more
expensive than sucrose while lactose, obtained from
whey, potentially is an inexpensive raw material.
The properties of the fatty acid esters of the glycosides of sorbitol, like those of lactitol and maltitol
are comparable with those of the known sucrose esters.
Sucrose has 8 hydroxyl groups which can be esterified,
whereas lactitol and maltitol contain 9 hydroxyl groups.
    Emulsifying agents can be prepared by means of our
method with values for the hydrophilic-lipophilic balance (HLB) varying from HLB 4 to HLB 16 by modifications of the percentage of monoester content, by the
type of fatty acid chosen and, particularly by the
choice of the polyol or polyol glycoside.   For instance, the stability of the fatty acid sorbitol esters
surprisingly is so high that, in the reaction with free
organic acids, only a relatively small amount of sorbitan compound is formed,whereas the reaction of sorbitol
lactate with fatty acids will result in a fairly high
conversion into anhydro-compounds.
    In the beginning of our research program, it was
expected that this last reaction especially would
result in a smooth esterification without anhydro-
formation, due to the better mutual solubility of the
reactants.
    To modify the properties of the polyol esters ob-
tained, we have treated them with several organic acids
including acetic,lactic, malic, citric and diacetyltar-

taric acids.
    In the reaction of the polyols with the fatty
acids, the molar ratio polyol/fatty acid may vary between 4:1 and 1:5. The fact is that in the reaction,
apart from monofatty acid esters, difatty acid and
higher esters also are being formed.   The monoester
content in the product can be varied by regulating the
mol proportion of polyol/fatty acid.   The fatty acids
used are natural fatty acids having 10-20 carbon atoms.
    The fatty acid soap may be added as an alkali
metal salt of fatty acids having 10-22 carbon atoms, or
prepared in situ by adding   alkali metal compounds to
the reaction mixture, such as alkali metal hydroxides
or salts of volatile organic acids.   The quantity of
soap can vary within wide limits, e.g.,10-80% by weight
of the quantity of polyol.
    The reaction velocity can be influenced signifi-
cantly by the choice of the reaction temperature and
quantity of soap, without the formation of too much of
the anhydro-compounds.   As the soap has to be eliminated from the reaction product after the reaction and
especially when the reaction product is to be used
directly without purification, the quantity of soap
chosen preferably will be as low as possible.   The
reactions usually are carried out under an inert gas
for the exclusion of oxygen.   The reaction products
can be purified by known methods, like the methods used
for the purification of sucrose esters of fatty acids.
In addition nonpurified products, containing some free
polyol and soap, may be used directly as emulsifiers.
    The yields of anhydro-compounds formed and the
residual polyol percentages are determined by gas-
chromatography.   For that purpose, the products are
saponified, the salts are removed by ion exchange
resins and the polyol mixture is acetylated, after
adding an internal standard.   The polyol glycosides
can be determined by means of quantitative, thin layer
chromatography.   The separated components are colored
by spraying with an aniline-diphenylamine-phosphoric
acid (adp) reagent (0.75 g aniline, 0.75 diphenylamine
in 50 ml ethanol with 5 ml 85% phosphoric acid) and
heated for 30 min at 110°C.   Concentrations are measured on the plate by means of a Vitatron TLD 100,
"flying spot" densitometer, referring them to standard
mixtures of known concentrations.
    The composition of the polyol esters also is deter-
mined by means of quantitative thin layer chromatography.   Monofatty acid esters, difatty acid esters
and higher esters are separated on silica gel, thin
layer plates by elution with a mixture of benzene, ether

and methanol (70, 35 and 7 parts by volume, respectively) for the determination of polyol fatty acid esters
and, by elution with a mixture of chloroform, acetic
acid, methanol and water (80, 10, 8 and 2 parts by
volume, respectively) for the determination of the glycoside esters.   The separated glycoside esters are
made visible by spraying with an adp reagent and treat-
ing them for 30 min at 110°C.   The retention times of
the various components of the maltitol and lactitol
fatty acid esters correspond fairly well with those of
sucrose fatty acid esters.   This is demonstrated in
Figure 2, a picture of a thin layer plate, showing the
separated components of the fatty acid esters of sucrose, lactitol and maltitol.   The similarity of the
elution patterns indicates a certain resemblance in
hydrophilic properties.

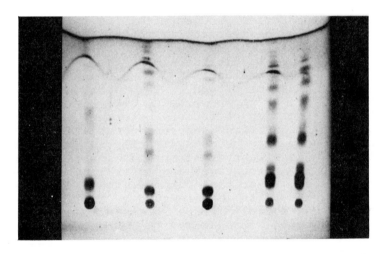

Figure 2.   Thin layer chromatography. TCL-2 microliter of a 2.5% solution in pyridine of: maltitol palmitate, (mol ratio: 1:1); lactitol palmitate,
(mol ratio: 0.4:1); lactitol palmitate, (mol ratio: 1:1); sucrose stearate, HLB
7; sucrose stearate, HLB 11.

Figure 3 gives data for the preparation of lactitol palmitate, as an illustration of the influence of
the molar ratio polyalcohol/fatty acid on the composition of the reaction product.   According to this pro-

cess a considerable variety of products with different properties can be prepared, which can be used in numerous applications, in analogy with the known emulsifiers on the basis of glycerol, polyglycerol, sorbitan and sucrose. For more experimental details of the preparation of these polyolesters, we can refer to the patents for which we have applied in several countries, e.g., U.S. patent 3.951.945, recently granted.

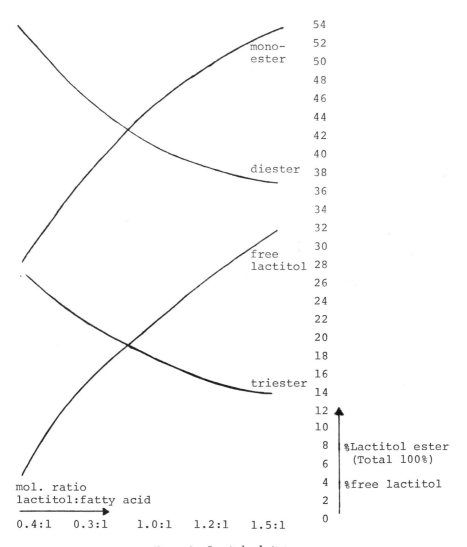

Figure 3. Lactitol palmitate

We will not discuss in this paper the various examples of the subsequent reactions of the polyol fatty acid esters to prepare additional derivatives by further esterification with food acids, such as lactic, citric and diacetyltartaric acids. The reactions are carried out at about 100°C at reduced pressures of 20-40 mm Hg.
The polyol esters have many interesting properties which make them useful as additives in foodstuffs, cosmetics and pharmaceutical preparations, detergents, etc. It is obvious that the functional properties of polyolesters, such as emulsification and stabilisation of emulsions, improvement of texture or consistency of foodstuffs, crystal modification, detergent activity etc.,are dependent largely on their chemical structure: type of polyol and fatty acid,mono- di- triester ratio, presence of soap and free polyols, etc. For each specific application, the chemical composition has to be optimized.
For several applications a certain analogy was found between the properties of the polyol esters and those of chemically related groups of emulsifiers. Confining ourselves to lactitol palmitate we observed certain properties analogous with sugar esters. For some applications, lactitol palmitate gave the best results, in other applications sugar esters were preferable. Some of the properties of lactitol palmitate are illustrated by the following experimental examples. In these examples the lactitol palmitate used was an unpurified product, direct from the reaction.
The emulsifier properties of lactitol palmitate were determined as follows:
Five g of the emulsifier was dispersed in 450 g water. Then 50 g soya oil was added and this mixture homogenized. The emulsions were poured into 100 ml calibrated cylinders and the emulsion stability was determined after 24 hours, by measuring the percentage of upperlayer of the emulsion. The results are shown in Table I.

Table I.   Emulsion stability data

Emulsifier	% upper phase after 24 hr
Lactitol palmitate	1 %
Sorbitan monolaurate	8 %
Sorbitan monopalmitate	37 %
Glycerol monostearate	6 %

The diacetyltartaric ester of sorbitol palmitate was found to be very effective in liquid coffee white-

Table II   Detergency data of polyol and sugar esters expressed as increase of reflectance after washing of the test fabric.

stain type	fabric type	no surfactant	ethoxylated tallow-fatty alcohol	lactitol palmitate	sugar ester HLB 7	sugar ester HLB 14
fat + soil stains	cotton	14.6	19.4	19.9	11.7	21.4
	PE cotton	13.4	23.4	23.8	13.1	26.2
food stains	cotton	7.8	4.5	17.3	15.6	15.9
	PE cotton	37.0	36.7	39.2	40.1	40.6
natural colours	cotton	25.2	22.9	22.4	22.3	24.4

ner formulations.   It offers good emulsifying and stabilizing properties and increases the freeze-thaw stability of the product.   The citric acid ester of sorbitol palmitate gave good results in several cake and
sponge cake formulations.   Lactitol palmitate and
several ionic derivatives of sorbitol palmitate showed
good antispattering properties in frying tests.   The
detergent activity of lactitol palmitate was determined by washing cotton and polyester-cotton test
fabrics with an international reference washing composition, in which the detergent component was replaced
either by lactitol palmitate or sugar esters.   The
water hardness was 180 ppm $CaCO_3$, the temperature 60°C.
The detergent activities were determined by measuring
the increase in reflectance of the test fabrics after
washing.   The data are summarized in Table II.
Lactitol palmitate showed very promising detergent
properties, which are in accordance with the results
of F. Scholnick, et al., Eastern Regional Research
Center, U.S.D.A., published in J.A.O.C.S. July 1975.
They have prepared lactitol palmitate by reesterification of methylpalmitate with lactitol in DMF.

Abstract

        Polyols are esterified directly with fatty acids
of edible fats in such a way that formation of anhydro-
polyols are minimized.   The reaction is carried out
at a relatively low temperature of about 150°C with
sodium soaps of the fatty acids as catalysts.   Examples of the esters made by this process include sorbitol, mannitol, lactitol and other polyols.   Additional
derivatives are made by further esterification with
organic acids such as citric and diacetyltartaric acid.
The polyol esters exhibit a surfactant character equivalent to the sucrose esters and find uses in similar
applications.

Biographic Notes

        J.A. van Velthuijsen, Drs. Manager of R & D.
Educated in The Hague and the Univ. of Amsterdam.
In 1965 joined Chemie Combinatie Amsterdam; Studies
of production of chemicals from sugars and polyols.
C.V. Chemie Combinatie Amsterdam C.C.A., Gorinchem,
Holland.

Organometallic Derivatives of Sucrose as Pesticides

R. C. POLLER and A. PARKIN

Chemistry Department, Queen Elizabeth College, University of London, Campden Hill Rd., London W8 7AH, England

Organotin pesticides,usually containing three tin-carbon bonds (R_3SnX), are unique in that biological activity disappears when the organic groups are removed from the metal. This process occurs on exposure to light and microorganisms (1,2) and these pesticides, therefore, present no threat to the environment. Their major disadvantages are high cost and lack of specificity so that prey as well as predator, may be attacked.

The aim of the present work was to examine the effect on biological activity of making a major change in the solubility of R_3SnX compounds by introducing a sucrose residue into the X group. This method of increasing solubility seemed attractive since glycoside formation commonly is used in nature to effect transport in tissue cells of otherwise insoluble materials.

Because of the nature of the functional groups present in carbohydrates, few of their organometallic derivatives have been reported. Organotin groups have been linked to cellulose by a variety of methods (3,4). In a few instances hydrogen atoms of hydroxyl groups in sugars have been replaced by organotin groups (5-7) but these alkoxy derivatives usually are sensitive to water. More recently, we have described some more stable compounds in which an organometallic group is joined to a sugar by a sulphur atom (8). None of these methods seemed appropriate to our purpose and we decided to prepare organotin compounds from sucrose hydrogen phthalate and sucrose hydrogen succinate, since these could be obtained by direct reactions between sucrose and phthalic or succinic anhydrides, as shown in Figure 1. The maleate was difficult to isolate so that later work was confined to the phthalate and succinate.

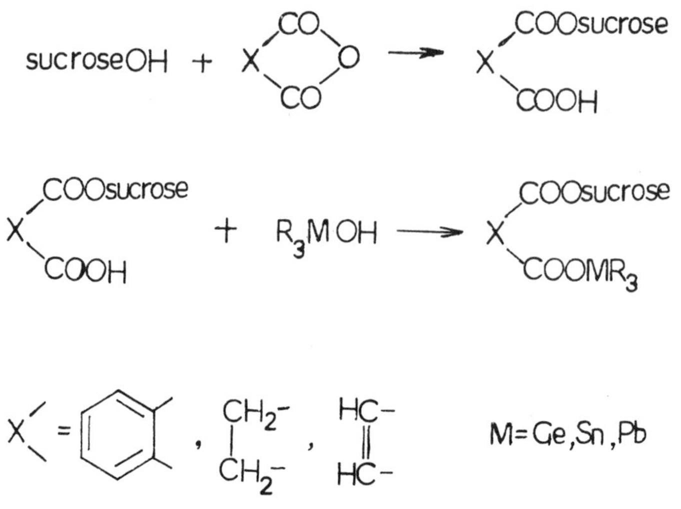

Figure 1

We were interested in commercial exploitation and chose to use free sucrose and accept that, in addition to variations in the extent of substitution, there would be complications due to regioisomerism. All of the biological tests were carried out with the crude products obtained from the above reactions. Before giving details of the results of these tests, reference will be made to some aspects of the work we have done on chemical characterisation of the sucrose esters.

With regard to the degree of substitution, two approaches have been used, i.e. separation of silylated derivatives by gas chromatography (glc) with mass spectrographic examination of the peaks and separation of acetyl derivatives by column chromatography. A typical glc trace is shown in Figure 2. From right to left, the first peak is due to solvent, the next sharp peak is octa(trimethylsilyl)sucrose and the silylated sucrose succinates then appear as a poorly resolved, composite peak. Hence, although we can acurately estimate the amount of unsubstituted sucrose, we are less certain of the relative amounts of mono- and di-ester. The second technique, whereby sucrose octaacetate and the various acetylated sucrose esters are separated by column chromatography, yields more reliable information. In Tables I and II the two methods are compared and we see that agreement is good in the case of the phthalate ester, less so for the succinate and that both products contain substantial amounts of free sucrose.

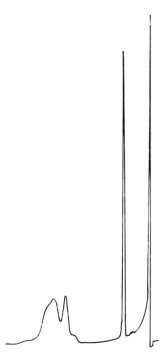

Figure 2. GLC trace of "sucrose succinate" after trimethylsilylation

Table I

Composition of Sucrose Phthalate

Method

	Separation of Acetates (%)	Separation of Me$_3$Si derivs. (%)
Free sucrose	36	35
Monoester	37	42
Diester	26	24
Higher esters	1	0

Table II

Composition of Sucrose Succinate

Method

	Separation of Acetates (%)	Separation of Me$_3$Si derivs. (%)
Free sucrose	22	32
Monoester	43⎤	
Diester	34⎦	65
Higher esters	1	3

We are not in a position to make any precise
statements about regioisomerism within the mono- and
diester classifications but information can be gleaned
from a somewhat speculative interpretation of the ^{13}C
nmr (nuclear magnetic resonance) spectra of these products.   Work in this area still is in progress and,
for example, our tentative conclusion regarding sucrose
phthalate is that substitution has occurred at the
6,6' and 4 positions in sucrose.
    It was of interest to include in our biological
tests a compound in which the sucrose residue was incorporated into an R group of an R$_3$SnX compound.   To
this end we utilised the series of reactions shown in
Figure 3, beginning with the sodium salt of the crude
sucrose phthalate.   Given the complexity of the starting material, the final bromo compound, of course, is
a mixture but its analysis, infra red (ir) and nmr
spectra correspond to the formula shown.

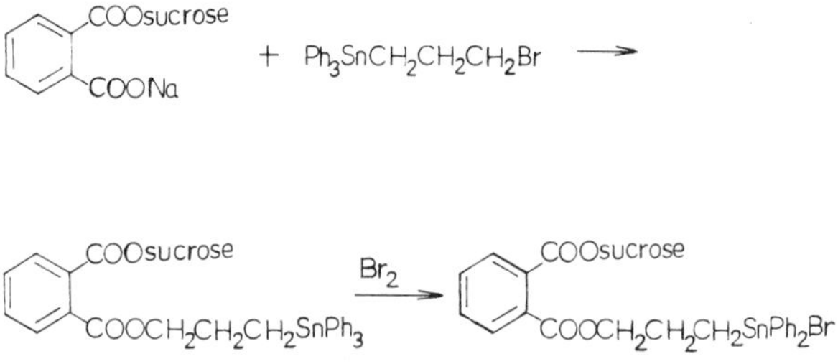

Figure 3

Turning now to the biological activity of these
compounds the results of tests against various organisms are summarized in Tables III - VI.

Table III   Activities against paint-destroying fungi

No. of fungi showing inhibition
of spore germination (max 16)

Compound	100 ppm	10 ppm	1 ppm
$Bu_3SnOCOC_6H_4COOsucrose$	16	16	6
$Ph_3SnOCOC_6H_4COOsucrose$	16	14	4
$(C_6H_{11})_3SnOCOC_6H_4COOsucrose$	3	0	0
$Bu_3SnOCOCH_2CH_2COOsucrose$	16	16	10
$Ph_3SnOCOCH_2CH_2COOsucrose$	16	15	9
$SucroseOCOC_6H_4COO(CH_2)_3-$ SnPh$_2$Br	16	7	0
$(Bu_3Sn)_2O$	13	13	12

Table IV   Activities against Enteromorpha: compounds
tested against Enteromorpha in sea water modified with
algal nutrients.

Compound	Concentration	
	1 ppm	0.1 ppm
$Bu_3SnOCOC_6H_4COOsucrose$	+	+
$Ph_3SnOCOC_6H_4COOsucrose$	+	+
$(C_6H_{11})_3SnOCOC_6H_4COOsucrose$	-	-
$Bu_3SnOCOCH_2CH_2COOsucrose$	+	+
$Ph_3SnOCOCH_2CH_2COOsucrose$	+	+
$SucroseOCOC_6H_4COO(CH_2)_3-$ SnPh$_2$Br	+	+

+=effective   -=not effective

Minimum concentration at which $(Bu_3Sn)_2O$ is effective
= 0.3 ppm

Table V.

Antibacterial properties of organotin compounds

Compound	Escherichia coli Conc^n (ppm)	Escherichia coli Inhib^n (%)	Micrococcus denitrificans Conc^n (ppm)	Micrococcus denitrificans Inhib^n (%)
$Ph_3SnOCOC_6H_4COO$sucrose	84	100	0.25	100
$Bu_3SnOCOC_6H_4COO$sucrose	-	-	1	100
Sucrose$OCOC_6H_4COO(CH_2)_3$-$SnPh_2Br$	83	23	1	100

Table VI

Molluscicidal Activity

Compounds tested against adult Biomphalaria glabrata. The LC_{50} assessed after a 120 h recovery period.

Compound	LC_{50} (ppm)
Triphenyllead Sucrose Phthalate	0.1-0.2
Triphenyltin Sucrose Phthalate	0.2
Triphenyllead Sucrose Succinate	0.1-0.2
Tricyclohexyltin Sucrose Phthalate	0.05
Tributyltin Sucrose Phthalate	0.1
Triphenyltin Sucrose Succinate	>0.2
Tributyltin Sucrose Succinate	0.075-0.1

These results show that the tributyltin and triphenyltin derivatives of sucrose are at least equal in biological activity to the commercially used tributyltin oxide and fluoride and are superior with regard to algicidal and fungicidal properties (9, 10). This is even more remarkable when we compare the tin contents, see Table VII. Except in the case of molluscicidal activity the tricyclohexyltin compounds are less effective and, in most tests, the less accessible compound with the sucrose residue in the R groups is inferior. Tests on the organolead compounds so far have been confined to molluscicidal activity.

Table VII

Tin content of biocidal organotin compounds

Compound	% Sn
Tributyltin oxide	39.8
Tributyltin fluoride	38.4
$Bu_3SnOCOC_6H_4COO$ sucrose	15.2
$Ph_3SnOCOC_6H_4COO$ sucrose	14.1

Tributyltin sucrose phthalate and succinate have been formulated into paints and subjected to larger scale tests. The fungicidal activity is retained and the compounds appear to give enhanced protection to weathered films of emulsion paint when compared with tributyltin oxide.

In the trialkyl and triphenyl compounds described above, the sucrose group is joined to tin via an

$$-\overset{\mid}{\underset{\mid}{Sn}}-OCO-\overset{\mid}{\underset{\mid}{C}}- \text{ group.}$$

Although the biological tests and other measurements indicate that this group is much more resistant to hydrolysis than might be expected, it, nevertheless, represents a point of structural weakness in the molecule. We therefore, have been examining other methods of joining organotin groups to sucrose, again confining ourselves to processes of potential commercial interest. The most effective of these is as follows:

$$R_3SnSCH_2COOMe + sucroseOH \rightarrow R_3SnSCH_2COOsucrose + MeOH$$

As before, the product likely is to be a mixture and we as yet do not have details of its complexity. The fungicidal properties of this type of compound are being evaluated. Preliminary results indicate that they are somewhat less active than the phthalates and succinates, but better than tributyltin oxide, on a weight for weight basis.

Acknowledgements

We thank the International Sugar Research Foundation, Inc., for supporting this work. We are grateful to the Paint Research Association for carrying out the tests against fungi and Enteromorpha, to Dr. P. Norris (Microbiology Department, Queen Elizabeth College) for carrying out the tests against bacteria and to Dr. J Duncan (Centre for Overseas Pest Research) for the molluscicidal screening.

Abstract

Methods for attaching organotin groups to sucrose
are discussed. Treatment of sucrose with phthalic
(or succinic) anhydride gave a mixture of sucrose hydrogen phthalates (succinates) which reacted with organotin hydroxides, (R_3SnOH R=alkyl or aryl), to give
organostannyl sucrose phthalates (succinates). The
corresponding organolead and organogermanium compounds
were prepared by similar methods. The biocidal activities of the organotin derivatives of sucrose were
found to be much higher than would be expected from
the tin content.

Literature Cited

1. Barnes, R.D., Bull, A. T., Poller, R.C., Pestic.
    Sci., 1973, 4, 305.
2. Chapman, A. H., Price, J. W., Int. Pest Control,
    1972, 14, 11.
3. Shostakovskii, M. F., Polyakov, A. I., Sinagin,
    V. V., Mirskov, R. G., Zh. Prikl. Khim., 1968,
    41, 2796.
4. Arcemova, Yu V., Vinnik, A. D., Zemlyanskii, N.W.,
    Rogevin, Z. A., Cellulose Chem. Technol., 1971,
    5, 319; (Chem. Abs., 1972, 76, 155815.)
5. David, S., and Thieffry, A., C.R. Hebd. Seances,
    Acad. Sci. Ser. C., 1974, 279, 1045.
6. Wagner, D., Verheyden, J. P. H., and Moffatt, J.G.
    J. Org. Chem., 1974, 39, 24.
7. Crowe, A. J., and Smith, P. J., J. Organomet. Chem.
    1976, 110, C59.
8. Husain, A. F., and Poller, R. C., J. Organomet.
    Chem., 1976, 118, C11.
9. Parkin, A., and Poller, R.C., British Patent Appl-
    ication No. 12168/76.
10. Parkin, A., and Poller, R.C., Pesticide Science,
    in press.

Biographic Notes

Robert C. Poller, Ph.D., Reader in Chem. Educated
at Univ. of Southampton. Chemist in the food industry; post-doctoral work at Birkbeck Coll.; organometallic chemistry and polymer stabilization. The
Chemistry Department, Queen Elizabeth Coll., Univ. of
London, Atkins Building, Campden Hill Road, London
W.8, England.

Discussion

Question: Mr. Kosaka, your sugar esters were produced using a solvent. What solvent did you use?
DMF?

Mr. Kosaka: Ryoto does not use DMF, but I do not
deny that the company is using some kind of solvent in
the process.

Question: Is it an edible solvent, and is it non-
toxic?

Mr. Kosaka: It is one of the safest solvents,
however, the products are solvent-free.

Question: Dr. Parker, have you done any experiments in making your sugar ester by the solventless
process using impure sugar rather than pure sucrose?
If so, what results did you get?

Dr. Parker: We have used impure sugar but not
molasses, which has too high an ash content. The main
reason for not doing so is that the sugar represents
a relatively minor cost constituent, and traditionally
we have used refined sugar. We do not have to, but
there is no particular advantage in not doing so.

Question: Monsieur Bobichon, could you divulge in
some general way how you managed to reduce the dimethylformamide to 5 ppm?

Mr. Bobichon: I cannot answer that in too much
detail - just by taking pains in the process, being
very careful. I do not think that the dimethylformamide is really such a bad material, and I agree with
what Dr.Hass said this morning about the inappropriate-

ness of zero tolerance for toxic materials.

Question: What is meant by a half-day diarrhea?

Mr. Bobichon: We measure the diarrhea of the animals twice a day by visual evaluation, and record the
total number of observations and the number of definite
diarrhea.

Question: Is that after 84 days of feeding?

Mr. Bobichon: After 84 days of feeding, we have
168 evaluations. As an example, we could have 14
positive diarrhea observations. Then we would record
14 as the number for the half-day diarrhea.

Question: How do the economics of the three surfactants compare with the petrochemically-derived products, on a cost/performance basis?

Dr. Parker: It is very difficult to answer for
other processes. Our own work was based on the fact
that we wished to obtain a sucrose-derived surfactant
which was indeed competitive with linear alkylbenzene
sulfonate-types of surfactants. Our product, as produced at the moment, is somewhat cheaper. We also
have a range of more highly purified products which are
more expensive, but are intended to be competitive with
non-ionic surfactants of the sorbitan ester-type. I
think processes using solvents inevitably must be somewhat more expensive, because of the need to reduce solvent residues to a minimum. We were not able to see
how we could do this, but maybe, Monsieur Bobichon,
could give his point of view.

Mr. Bobichon: It is difficult for a surfactant to
be competitive primarily because of the pH stability.
We tried many times to find applications in surfactants
for our sucroglyceride, but we failed because of the
instability at alkaline pH.

Question: Dr. Parker, will you comment on that,
please?

Dr. Parker: We found that the sucroglycerides are
stable over a pH range from 6 to 12. Obviously, they
are not stable under acidic conditions,and this constitutes a limitation. But, the number of applications
requiring acid-stable surfactants are relatively few.
We are not worried by this, although we would like to

have one for these particular applications.

Question: Mr. Kosaka, would you like to say anything about the economics of your product?

Mr. Kosaka: The recovery of solvent in the Ryoto
process is greater than 99%.   Thus, the cost of solvent is negligible, and our products are solvent-free.

Question: Mr. Kosaka, would you elaborate on the
bakery uses of sucrose esters in Japan?

Mr. Kosaka: The use in cakes is most popular.
Japanese bakers are using a mixture of sugar ester and
monoglyceride in bread.

Question: For high protein breads, or just regular wheat bread?

Mr. Kosaka: Just regular bread.

Question: Is that application as an anti-staling
agent?

Mr. Kosaka: Yes, as an anti-staling agent,mostly.

Question: How difficult is it to control the residual methanol content,and the toxicity of this methanol?   If there is any problem, have you thought of the
use of the ethyl esters for the transesterification?

Mr. Kosaka: We thought about using ethyl ester,
but its cost is higher than that of methyl ester.   We
cannot find methanol in our products.

Dr. Hass: Minute traces of methanol are not toxic.
The pectins which are used for making jellies have car-
bomethoxy groups which are split off in the digestive
system to form traces of methanol, and they do not hurt
anybody.

Dr. Hickson: With regard to the ethanol-methanol
question,do not forget that ethanol boils at a slightly
higher temperature and requires a little more heat to
take it off.   The reaction calls for the distillation
of methanol to provide the driving force to create the
sugar ester.   That is a major reason why the methyl
ester was chosen.

Question: Professor Seib, what is the potential

total volume of the surfactant market, in baking, in
the United States, alone?

Professor Seib: I tried to intimate what the total market might be by tabulating the poundage of the
various baked foods in the United States. If 70% of
that poundage is bread and rolls, one could estimate
about 90,000 lb/day of surfactant for that part of the
market. The next largest use probably would be cake.
I calculated another 10,000 lb of ester/day in that
market. That is based on a total of 1.4 billion lb
of cake annually, which contains 3 to 6 % surfactant in
the emulsified shortenings which are used in cake manufacturing. I think that, certainly, is not the total,
but you can see how quickly you drop off in magnitude
from 14 - 17 billion lb of bread and rolls down to 1.4
billion lb of cake per year. I think that a potential
market of a 100,000 lb/day of sugar esters in the baking industry is a fairly good approximation. Mr.Beatty,
of Continental Baking, has just mentioned that 450,000
tons of bread flour/year is used by his company alone.

Question: Dr. van Velthuijsen, is the price of
the sugar alcohols much higher than the price of sugar?
How does this affect the total economics of the final
product?

Dr. van Velthuijsen: The price of the sugar alcohols is about double or triple the price of sugar. Of
course, this enters in the costing of the product.
Even so, the price of the crude reaction mixture is in
the same price level as the surfactants, such as the
glycerol monostearates and stearoyl lactylates, now
available in the foods market.

Question: Dr. Poller, are there any plans to go
forward with field testing of any of the pesticides
you mentioned?

Dr. Poller: As I indicated in my talk, this has
been done for applications in paints. Large-scale
tests have been carried out. For more general uses --
for instance, in agriculture -- we are at present just
coming to the end of synthesizing large-scale batches
for exactly this purpose.

Professor Vlitos: The screening of compounds of
this type is rather difficult. The syntheses usually
are done in a university laboratory, usually in the
chemistry department, and the screening has to be done

by biological groups. Very often they are busy doing
their own types of research. If there is any weakness
in the approach of the International Sugar Research
Foundation it is that it has been found to be very
difficult to screen the usefulness of the compounds
that have been synthesized in the research programs.
There are many, many compounds, which may have, who
knows how many applications. We just have not had a
chance to really look at them all in the right test
systems. I think Dr. Poller's work indicates: first,
here is some interesting chemistry; and second, his
compounds have a use which already has been demonstrated. Now, it is up to somebody to synthesize them and
sell them. That is another problem.

Question: Dr. Poller are these materials soluble
enough in water to use them, or are they completely
solvent-soluble?

Dr. Poller: The solubilities of the standard,
organotin pesticides currently in use are low -- some
of them very low, indeed -- in parts per million. By
putting on the sucrose group, the solubility goes up
around a thousandfold, as a rough estimate, but we
have no precise figures.
This is an interesting question. Is this simply
a solubility effect, or is it something else? I conclude it is not simply a solubility effect because, at
levels where the tributyl tin oxide is soluble, the
activities of the sucrose compounds are higher. In
the best tests, the sucrose compound is three times
more effective with around a third the amount of tin in
it. Most of the tests were carried out at concentrations where both the sucrose compounds and the reference organotin pesticides were completely dissolved
and so the increased activities cannot simply be attributed to solubility effects.

Question: Would the high solubility cause difficulties in formulation?

Dr. Poller: Not necessarily. The organotin esters can be hydrolysed and it is possible that, after
transport in living tissue as the organotin sucrose
ester, hydrolysis occurs and the insoluble organotin
oxide is then deposited.

Question: What kind of breakdown products do
these sucrose tin compounds have on biodegradation?

Dr. Poller: We do not know. The sucrose tin derivatives are very new, and they still are being tested. However, we have done some work on biodegradation of triphenyltin pesticides. This was done by synthesizing ^{14}C-labelled pesticides, putting them into soil under various conditions, and then collecting the radioactively labelled carbon dioxide. The organic groups come off in a stepwise fashion and break down completely, to carbon dioxide. Only inorganic tin remains.

Question: Is it definitely inorganic tin, and not some tin-alkyl derivative such as a methyl tin compound?

Dr. Poller: Yes. A ^{14}C-labelled, radiochemically and chemically pure compound is added to soil, and the radioactivity of the $^{14}CO_2$ evolved is measured carefully. When all of the ^{14}C label appears as carbon dioxide there is no doubt that the compound is broken down completely. This is the sort of evidence we have that the triphenyltin compounds are converted to inorganic tin.

Question: Are you certain that an organotin compound could not find its way into the food chain some other way, possibly through solubilization processes, transport processes, and so forth?

Dr. Poller: It is impossible to be certain, but the evidence we have suggests that organotin pesticides are converted to inorganic tin on exposure to light and microorganisms. The wide use of tin-plated cans as food containers and much other evidence indicates that inorganic tin compounds are not toxic.

Dr. Hass: It was the idea of one of our librarians at Sugar Research Foundation that the general approach of combining sucrose with pesticides might be effective. We first tried it in a project at Battelle from which we never got anything that was as good as existing pesticides. I am delighted that you are doing it, and that we simply tried the wrong compounds.

Surface Coatings and Other Esters

Introduction

JOHN MORTON

Redpath Sugars Ltd., P.O. Box 490, Montreal, H3C 2T5 Canada

The first session of this Symposium brought us up to date in several areas of basic sucrochemistry. We heard that an immense number of esters have been identified and their properties studied, yet the potential for future research appears almost limitless and this work no doubt will provide papers of interest for many years to come.

The second session gave us a preview of some commercial ventures in the application of sucro-esters. Examples were cited in calf feeding, the baking industry, and one of the papers indicated some potential for sucrose esters in the field of pesticides.

This third session will deal with some further applications in the field of surface coatings, in which there would seem to be great opportunities for the use of sucrose esters, both as the vehicle in film forming and as a modifier of the film characteristics. One is aware of the technical promise of much of the early work on coatings vehicles, but the commercial possibilities in this regard have yet to be exploited fully. In the application of special esters in plasticized films and as film modifiers, there has been more success, as documented in the following papers.

12

Prospects and Potential for Commercial Production and Utilization of Sucrose Fatty Acid Esters

EDWARD G. BOBALEK

University of Maine, Orono, Maine 04473

The stage was Case Institute of Technology because it has very close relations to the surface coatings industry; the time, the middle 1950's; the author, Dr. Henry B. Hass, a grand provocateur, and Dr. John L. Hickson, prompter and interpretor, both of the Sugar Research Founation; the plot "what can be done with sugar esters, especially oil-fatty acid esters, using the oil acids that commonly occur in soft, semi- or hard drying oils?".

The players in the cast included myself, Dr. Thomas J. Walsh in Chemical Engineering, and several graduate students who broke their teeth on the bit of sugar chemistry, particularly Drs. M.T. Chiang and C.C. Lee, W.J. Collings, Alfredo Causa, Alfredo deMendoza and George Kapo. Much of the contribution of the work which ensued really comes from their efforts and their aptitudes as they grew in this science.

At that peculiar time, the coatings industry, particularly in its vehicle aspects, was less than 10 years away from the open-fired, varnish kettles. It was substantially based on natural drying oils. It was in a ferment of revolution trying to cope with the problems of material shortages. As in all conservative industries, the approach to coping is the same. That is, a request to supply something that is remarkable, but that does not force changes which are not absolutely necessary.

The goals were simple. That is, not to begin at the beginning, but to study what might be found in established markets. Table I indicates some of the primitive targets.

Table I. Possible Sucrose Ester Applications

1. Printing inks.
2. Wood Sealers
3. Varnishes and Paint Vehicles
    Oleoresinous Blends
    Diisocyanate Modifications
4. Emulsion Vehicles

Drying oils and drying oil products already were critical in these particular utilizations. They represented, at that time, a large segment of the consumer products market. This region also was in a ferment of
development. The chemical process industry had come
out of World War II with a capability for a multitude
of by-products for which it had no market. The most
receptive domain where they could get a friendly testing and be accepted, if possible, was thought to be in
the paint, varnish, printing inks and related industries. Figure 1 gives something of the scenario.

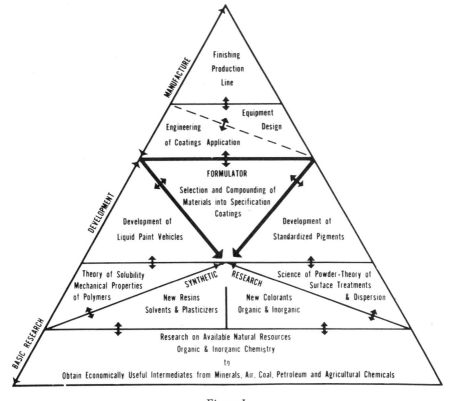

Figure 1

At that time the Division of Paint, Varnish,Print-
ing inks and Plastics probably enrolled about 15-20
percent of the American Chemical Society directly or
indirectly affiliated as providers of the final products, or vendors of intermediates.   In addition,
about 80 percent of them were showing up at the national meetings.

The point to be made is that the Sugar Research
Foundation had housed itself mainly in the base of the
pyramid, in basic organic chemistry.   Much of the work
it had been supporting had been down in this domain.
Where they wanted to get was up in the markets at the
peak of the pyramid.

At that time in particular, and still today, there
was a lot of brilliant research in the technology of
the products mentioned.   But, the key figure was the
"formulator", who constituted the image of the company
making the plastics, printing inks, paints, etc.   The
formulator occasionally could transcend the research to
solidify an invention.   Needless to mention very little research from satellite suppliers or the users
could enter the system except through his coordinating
position.   If something was going to be proof-tested
and brought from the research base up to the production peak, it would have to go through the formulator
in such a condition that he could adapt it quickly
with a minimum of pain, make his evaluation and, hopefully, push it further up the pyramid.

The domain that Case was supposed to take care of
was to escort some esters from basic research to the
second stage, by applying industrial process chemistry.
Figure 2, presents our target.

Structure of Hexalinoleate Ester of Sucrose.

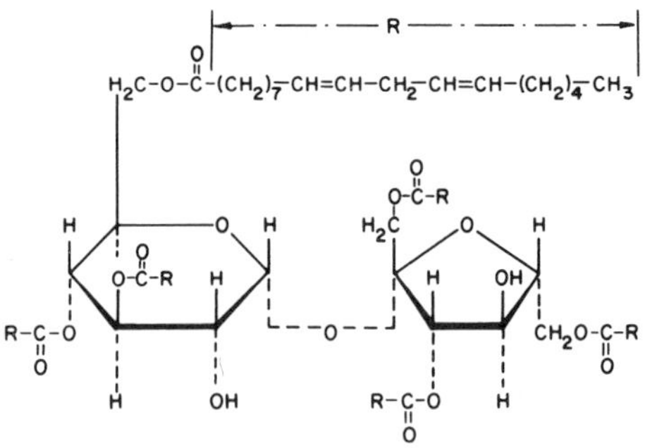

Reaction Stoichiometry is

$$C_{12}O_{11}H_{22} + 6 CH_3OOR \longrightarrow C_{12}O_{11}H_{16}(OOCR)_6 + 6 CH_3OH$$

Figure 2

There was relatively little useful sucrochemical
literature to go on at that time. The chemical literature itself was extensive, but not too adaptable to
the limited skills of chemical engineers. The only
sugar ester process that we really could get off the
bench was the transesterification process in a homogeneous solvent medium, which then was in a pending patent
by Dr. Hass and Dr. Lloyd Osipow. We came closest,
actually, to accomplishing the goal with this particular reaction. We do not claim that this reaction, as
first studied, will turn out to be the ultimate, the
best, and the final one. Since that time, an extensive monograph has been put together on the patent literature emerging since the mid-'50s which gives many
alternatives. We have learned, one can get away from
solvents by going through heterophase polymerization,
mass transfer exchanges between phases, use of different kinds of solvents, solid suspensions, liquid media,
etc. Yet those involved procedures, insofar as I have
been able to determine, and there has been no new dis-
closure in this Symposium, have never been put to the
critical test by the chemical engineers, even on a
bench scale.

Figure 2 illustrates the point. Before the reaction selected could get off the bench, it was necessary
to close the material balance. This means that repe-
titively, in at least a dozen tries, one measures the
percentages coming out in the products, what is going
up the stack, and what is being discarded in tars, and
whether this distribution is consistent.

The first reported objective then was that we
should come close to making a consistent material balance at a laboratory bench. This is a precursor step
on the way to a pilot plant. I will summarize hastily
some of the necessary, even if not entirely sufficient,
conditions.

First, the system has to be very anhydrous, thus a
feed stock preparation is essential. Secondly, one
must achieve a totally homogeneous reaction. The catalyst must be in solution or seeming so. The sugar must
be totally in solution. All of the intermediate products and final products must remain in solution. If
they do not, one runs into troubles with reactor problems. All reactors foam, spatter, mist and so forth.
They have hot zones and cold zones, crystallization
areas and charing areas. Maintaining beneficial color
and avoiding by-products usually is very difficult if
one must try to manipulate heterophase, suspension reactions. One demand is that there be no preperoxida-
tion or oxidation of the methyl esters. However,

there were techniques of stock preparation which
negated any oxidation that had occurred.   One wants
also a solvent which, in addition to being a good solvent for the whole reaction system, has a nice cleavage
possibility under distillation conditions for methanol.
There was some time before the sophisticated solubility
parameter concepts emerged, so that there was much tri-
al and error.   A good deal more can be done now on a
scientific basis.

The reaction rate has to be brought to reasonable
levels, hopefully to be achieved in 3-4 h.   Obviously,
this meant that the methyl esters had to be in surplus.
Sometimes at a ratio as high as 4, 6, or even 8 to 1
depending on what speed is wanted in order to produce
the desired higher esters.   That put demands on an
extraction, separation and recyle process.   The one
inherited from the Hass-Osipow patent, a silica gel
absorption technic, was not suitable because of its in-
convenience and the resulting side reactions which nul-
lified some of the benefits attainable by care in man-
aging the reaction.

Fortunately, it was found that the spectrum of
sucrose esters differed quite sharply in their temperature coefficients of solubility in methanol.   A simple
methanol extraction at varied temperatures could effect
reasonably good fractionations without degrading the
product.

Last, but not least, there could be translated a
closed material balance from the bench to the pilot
plant where chemical analyses of the progress of the
reaction seemed to coincide with the yields of the
methanol stripped off.   If one proceeded at low enough
temperatures, at high enough rates, one could minimize
decomposition of the solvent.   This was abetted particularly by sparging an inert atmosphere under reduced
pressures.

The reactor system is shown in Figure 3.   The
process starts with a stirred tank reactor with a good
reflex condenser and a good stripping still for the
methanol.   The discharge would go into a still for
stripping off the DMF.   The DMF-free reaction mix,
high in methyl esters would go over to a solvent extraction column.   Each extract, then would be purged again
of methanol, and sent back for recycle through the
system.

Figure 4 indicates something of the complexity of
the extraction system, which then was thought to be a
complex thing.   But, it was very primitive compared to
modern technology as it has emerged in the petroleum
industry.

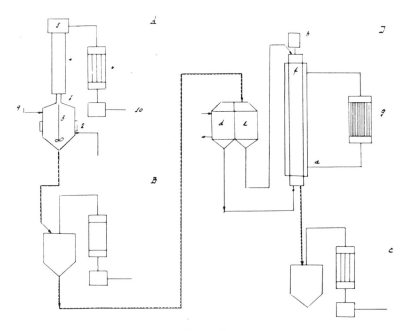

Figure 3

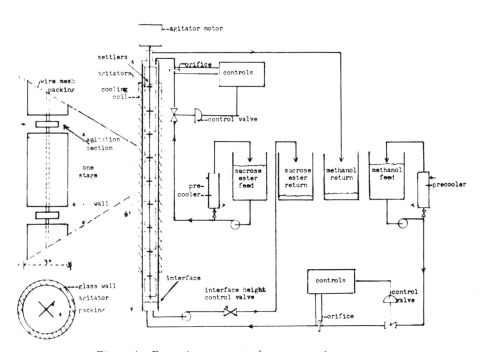

Figure 4. Extraction apparatus for sucrose ester process

    Now it becomes necessary to develop the ammunition
to corner the formulator and show him that use of the
new vehicle would lead to relatively few hazards when
applied in the practices, as he could understand them,
to convert this intermediate into some of his regular
coatings products.   Some of these data are tabulated
in Table II, which also illustrates a failure.   The
target, had been a degree of esterification (D.E.) of 8
or close to it.   It never was achieved; the highest
level achieved under the best conditions, keeping
within the restrictions of planned practicality, was
about 7.5 D.E.   Fortunately, it was discovered that,
to get adequate drying oil products, such a high DE
is unnecessary.   An approximate 5.5 D.E. was adequate.
Any further gain in D.E. in terms of coatings benefits
attained, was not that great.   Possibly this was because the D.E. began to be overwhelmed by the iodine
value (the degree of unsaturation) of the fatty acids
chosen.

Table II.

EFFECT   OF   DEGREE   OF   ESTERIFICATION   AND   IODINE   VALUE
ON   DRYING   RATE

DEGREE OF ESTERIFICATION	IODINE VALUE	DRY TO TOUCH(HR)	DRIED HARD(HR)	TACK FREE(HR)
6.1	191	2	8	10
6.1	175	2	10	14
6.1	148	3	16	20
6.3	152	3	14	20
6.1	148	3	16	20
5.6	153	3	16	22

FILM   THICKNESS – I MIL – CAST ON TIN AT 100 % SOLIDS

    In this light, the goal of a D.E. of 8, which may
be important for some esters, was abandoned for the
drying oils.   As a matter of fact, there evolved later
reasons to believe that some residual hydroxyl functionality is an important advantage to these products, when
used in conventional drying oil type applications.
Table III briefly places the situation into a context
of where the sucrose esters stand compared to the conventional products.   These are the physical properties
of the films.   Generally, the higher the number, the

better is the property, except for drying rates. The data compared here are for a sucrose hexalinoleate; a sucrose ester made by the most advanced techniques of the time, using entirely the methyl ester fraction from tall oil that corresponded in iodine number to pure linoleic acid, but it was not pure linoleic acid.

Table III.

COMPARISON OF FILM PROPERTIES

PROPERTY	LONG OIL ALKYD RESIN	SUCROSE HEXA- LINOLEATE	LINSEED OIL
TENSILE STRENGTH (PSI)	1450	730	180
ELONGATION (%)	110	80	95
MICROKNIFE HARDNESS	350	325	275
MICROKNIFE ADHESION	176	164	166
DRYING (HRS) — 40 % SOLIDS			
DRY TO TOUCH	0.8	1.0	1.5
DRIED HARD	1.0	1.5	2.5
TACK FREE	3.0	2.0	4.0
1 % ALKALI RESISTANCE	BLISTER-3 HR	BLISTER-2 HR	FAIL-10 MIN

One can notice that the sugar ester is closer in properties to the architectural, long oil alkyds, at that time, the most commonly used exterior paints in hard finishes, than to linseed oil itself, which was the common house paint base. However, in most fluidity properties and otherwise, it resembled a linseed oil. So one concludes the hexalinoleate looks essentially attractive.

Table IV.

OIL ABSORPTION RATIO FOR VARIOUS PIGMENTS

PIGMENT	OIL ABSORPTION RATIO SUCROSE HEXALINOLEATE / LINSEED OIL
TITANIUM DIOXIDE	0.37
PHTHALOCYANINE GREEN	0.81
MILORI BLUE	1.08
TOLUIDINE RED	0.80
LAMPBLACK	0.93

Table IV illustrates a feature that perhaps is the
most important property of all.   It tends to be ignored most often by people who have not had to live with
formulators of compounded products.   These materials
have to be blended with a variety of minerals,pigments,
fillers, and additives of all sorts.   Usually the index of quality is the minimum amount of oil that can
disperse, or turn into putty, a standardized quantity
of pigment.   This is true across the whole spectrum
of such formulated products from printing inks to
mastics.

In only one instance, with one pigment, Milori
blue, which people who know the art recognize as a very
miserable customer, did the sucrose ester appear a little worse than linseed oil.   In most cases, it took
less of the sucrose ester to effect this compositing
than of the linseed oil, the common grinding medium for
these dispersions.

These tables summarize only the main points.
There are weathering and many other tests that show the
sucrose esters in general are equal or superior to
traditional oleoresinous vehicles with which sucrose
esters hoped most to compete.

Amidst the course of this work an unusual observation was made during some precision studies on oxygen
absorption (Figure 5).   Those familiar with drying oil
science know that many side reactions occur in the
hardening of an oil in an oxidizing environment.   Some
of the reactions add oxygen, utilizing it in effect to
polymerize the oil, others decompose the oil.   There
always are weight losses being offset by weight gains
and the real index of quality is the net change.

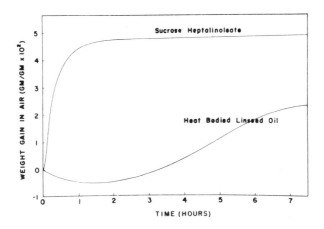

*Figure 5.   Oxidation rate of sucrose heptalinoleate vs. bodied
linseed oil*

Nearly always the sucrose esters, at D.E.'s of 5
or better seem to have a unique behavior, even at comparable iodine values.   It seemed that they peroxidiz-
ed and gained weight with minimal decomposition, compared to most natural oils.   Whether this was a condition entirely due to the greater purity of this ester
compared to natural oils, or whether some features of
structural chemistry were involved in it, must be left
to future investigators.

This observation did promote great interest.   Two
big competitive systems at that time were essentially,
vinylated oleoresinous materials, like styrenated alkyds and oils, and the emerging, emulsion vehicles like
SBR latices.

The sucrose esters could be oxygenated and the
oxygenated products were essentially stable peroxides.
The peroxides were stable even if dispersed in water
and could be dispersed with the help of their own, lower esters.   One could polymerize them with styrene,
styrene-butadiene, or vinyl acetate.   We did not try
hard to achieve all the possibilities, but there were
some indications that through this post-oxidation route,

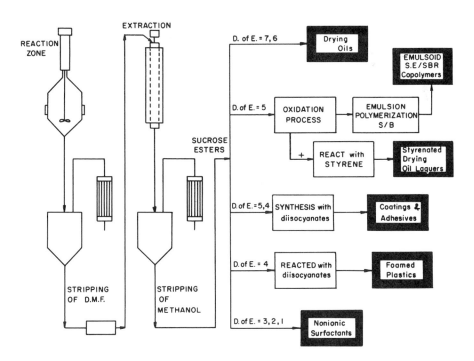

Figure 6.   Flow sheet from process to products

the sucrose esters could be entered directly either into styrenated,solvent resins(which dry in seconds,or in
minutes, compared to hours--really special kinds of latex) or even into emulsion technologies.

Figure 6 is a summary of where the study seems to
have left the prospects in about 1960. A process was
desired that was clean, environmentally closed, and
which could be operated in commodity quantities, to
provide various degrees of esterification of the sucrose esters. These were the outlets that were tested
and proven feasible to some degree, but for which there
was no extensive development to plant processes.

It was believed that the process would require
such heavy capital investment in whatever plant was
necessary, that it would have to make large volumes of
products, and should be flexible enough to produce the
variety of product lines that, at that time, were emerging in competition with the natural products.

Table V lists the obvious potentials as of 1960
when everything looked very good. Here it is 1976,
and nothing has happened. Perhaps a word is in order
on what, in retrospect, I think were some of the problems.

Table V.   1960 Potentials for Commercialization

    1.   Chemical research data exist
    2.   Pilot process design validated
    3.   Product quality is excellent
    4.   Raw material costs favorable
    5.   Marketing targets identified

    In 1966, the process we proposed (Table VI) was
too complex to fit the technology of the varnish plants
of that era; recycle, reflux, extraction,fractionation,
etc. The process costs were uncertain. The marketing futures were uncertain. The only thing that was
certain was that the use of the traditional natural and
refined drying oils was waning. New materials were
coming up. The new materials were coming less and less
out of the varnish plants of the formulator industry,
and more and more out of the chemical process industry.
It was not at all certain whether, if the chemical process industries, who were now vendors, should capture
all the vehicle manufactures, would expand also into
the final products.

Table VI.   1966 Resistances to Commercialization

1. Process complexity unacceptable
2. Process costs uncertain
3. Marketing futures uncertain
4. Biodegradation potential unwanted
5. R & D shifts to defensive posture

Biodegradation was a curse at that time.   Anything that biodegraded was regarded as not durable.
Mercury and other fungistats were being forbidden,
another direction that seemed likely to lead to abandonment of all oleoresinous products.
     On top of that, R & D shifted to a defensive pos-
ture.   All the Chairmen of the Board and Boards of
Directors of vendors and surface coatings companies
alike,concluded there was too much spending on research.
The results were not being used.   The emphasis was on
quick solutions to production problems.   In a period
of less than three or four years, a combined research
staff of 4,000 or so, productive scientists dwindled
to 600 or 700, or less over this whole industry.
     Perhaps I am inclined to put reason number five
as the main factor why the remaining problems were not
solved, and sugar esters were unable to be moved from
the formulators' corner up into commercialization.

Table VII.   1976 Potentials for Commercialization

1. Better chemical research data added
2. Engineering advances facilitate process
        design
3. EPA/OSHA etc. make closed process systems
        necessary
4. Material/energy needs favor renewable
        resources
5. New industrial structures open new markets
6. Renaissance of R & D

In 1976, one can see certain things that did not
exist, as shown in Table VII.   Generally speaking,
some excellent chemical research has continued and
better chemical research data on the synthesis of the
esters has been acquired.   In fact, it is possible
that the chemist of 1976 is becoming even slightly
aware of material and energy balances.   Thus, there is
less chance he will be delivering half-fried reactions
to the chemical engineer.   One can hope this now will
become habitual.

Secondly, engineering design has jumped eons in terms of new processes, particularly extraction, distillation, fractionation, stock preparation, etc. That is, processes that used to be cumbersome still may be capital intensive, but they can be fast.

On top of that, EPA, OSHA, etc., probably will never tolerate again the appearance of the kind of varnish plants that provided the low cost competition which constituted the barrier to setting up this new kind of system in 1960. The industry will be forced to go to closed systems. This sugar ester process is, should be, and can be an effluent closed system, particularly if one employs modern chemical engineering equipment, such as continuous reactors, film reactors, etc., for which this batch data is adequate to justify the first levels of design.

The raw material picture is improving. For example, in the paper industry kraft pulping now has displaced virtually all except mechanical pulping. The product stream of fatty acid by-products of kraft pulping today is about 6 or 7 times what it was in 1960. Astoundingly, the fraction that would best fit oleoresinous vehicles and sucrose esters is the one in least demand for competitive purposes. Best of all, these kinds of fatty acid wastes are being generated either from wood processing or food processing. One also gets close to many composite materials technologies like pressed-wood, fiberboard, and plywood, where all of these could feed back their own sugar esters as necessary intermediates.

In addition, these natural products industries; pulp, food processing and the like, come the closest that exist today to being self-sufficient in energy, on the basis of their own wastes.

One anticipates the overall generation of new product lines from natural products industries as closed situations. Here, many of the disadvantages now encountered with petrochemicals, and anticipated delays in the evolution of coal chemicals, may provide a favorable new opportunity for sucrochemicals that did not exist before 1970.

Finally,it seems important to point to the renaissance of R & D. This Sucrochemical Symposium is not only a symptom, but a symbol that finally the R & D situation has been turned around. Once again we may have to do some hard technological development for the funding is beginning again to emerge. We may again see an era when the research laboratories of the '60s will reappear in the coatings products industry.

Abstract

Laboratory and pilot plant data are reviewed critically to establish design guidelines for production of sucrose ester drying oils using process systems which are environmentally acceptable and can be economically feasible, providing that a limited number of standard products can be marketed in sufficient volume to offset high capital equipment costs needed to assure reliable product quality and efficient energy utilizaiton. Raw material supply and costs are very favorable. Sucrose esters of unsaturated fatty acids can be used directly to replace natural drying oils in traditional paints and varnishes, and also as major intermediates for a variety of copolymers or of heterophase emulsoid or organosol compositions that are potentially useful in design of coatings, plastics, adhesives or sealants. Increasing costs of competitive materials of petrochemical origin, and intensification of environmental and toxicity standards for consumer products, can make both direct and intermediate usages of selected sucrose esters more attractive today than seemed probable about 20 years ago. Some possible directions where product design research should be renewed are discussed.

Professor Edward G. Bobalek, Ph.D., Chairman, Chem. Eng. Dept. Educated at St. Mary's Coll. (Minnesota), the Creighton Univ. and Indiana Univ. Industrial chem. eng. at Dow Chemical Co., and Dir. Polymer Res. & Production, the Arco Co.; Case Institute of Technology 1948-1963, leaving as Prof. of Chem. Eng. to become Gottesman Research Prof. and Chairman of the Chem. Eng. Dept. Univ. of Maine at Orono. About 120 research papers. Dept. of Chem. Eng., Univ. of Maine, Jenness Hall, Orono, Maine 04472 U.S.A.

13

Surface Coating Sucrose Resin Developments

R. N. FAULKNER

Paint Research Assoc., Waldegrave Rd., Teddington, Middlesex, TW11 8LD, England

The potentials of higher substituted sucrose drying oil esters were demonstrated some twenty years ago
(1). Since then considerable effort has been expended
towards their commercialization. Sucrose linoleates,
with an average degree of substitution of 6 or 7, were
most promising, being superior to linseed oil and similar to alkyds with respect to drying and film forming
properties and alkali resistance. Esters with lower
degrees of substitution showed reduced performance, although the tetraesters still were claimed to be better
than linseed oil.

Substitution of up to substantially 5 sucrose
hydroxyl groups (presumably the 3 primary and 2 reactive secondary groups) is achieved fairly readily by reaction of a 1:5 mole ratio of sucrose and methyl ester
using the standard transesterification method. This
involves the use of an aprotic solvent, and "bone-dry"
potassium carbonate as catalyst at about 100°C. Synthesis of hexa/heptaesters is more difficult and it becomes necessary to use a 100% excess of methyl esters
to achieve a 90% conversion of sucrose. The unreacted
esters are recovered by chromatography or methanol extraction.

The work described in this paper aimed to synthesise high performance resins from the more readily prepared, lower sucrose esters with up to 5 drying oil substituents, through reaction of residual hydroxyl groups
with suitable cross-linking systems. Direct esterification with acid catalysts or high temperatures as are
used in alkyd resin manufacture clearly are unsuitable.
Reactions, therefore, were chosen that are effective at
temperatures which avoid decomposition of the reactants
and can be carried out using conventional, resin-making
techniques.

Experimental

Synthesis and Characterisation of Sucrose Esters.
Partial sucrose esters were used as starting materials
having the general formula $\underline{Suc}(OR)_n(OH)_{8-n}$ where; n =
3 to 5, $\underline{Suc}$ represents the sucrose skeleton, and R an
acyl radical. The R groups were derived from unsaturated fatty acids such as linseed and tung (drying oil
type) soya and dehydrated castor oil (semi-drying oil
type) and olive (non-drying oil type) or saturated
fatty acids such as lauric.
    Synthesis was carried out on a laboratory (up to
500g) scale by transesterifying sucrose at 90° - 100^oC
and 30 mm under nitrogen with three to five molar
proportions of methyl fatty esters (acid value preferably less than 1 mg KOH/g) in dimethylsulphoxide (DMSO)
(about 30 - 35% of the total reactants) and about 0.5
to preferably not greater than 2% dry weight of anhydrous potassium carbonate on the weight of methyl esters. Low amounts of catalyst and low acid values reduce contamination of the product with potassium soap.
    The reaction was followed by determining, by gas
liquid chromatography, the amount of methanol produced
by transesterification. In all cases 85 to (usually)
90% extent of reaction was noted. The reaction mixture was cooled, filtered through muslin, and crude
sucrose ester, containing unreacted methyl esters and
potassium soaps, was recovered in about 95% yield by
vacuum distillation of DMSO at 1 mm pressure. In certain cases, the amount of unreacted methyl esters was
determined by molecular, pot distillation.
    Purification to remove potassium soaps was effect-
ed by dissolving the crude product in about twice the
weight of methyl ethyl ketone and washing with 0.1 N
hydrochloric acid (in an amount corresponding to 100%
excess of the quantity required for neutralisation),
followed by dilute brine and finally water. The recovered product was pale in color after distillation
of solvent. A low percentage loss was noted at this
stage. This product was further upgraded, in the case
of the "tetra" and "penta" esters, by washing with
about 2 to 3 times its weight of methanol to remove
soluble mono- and disubstituted sucrose esters and
fatty acids. (Extraction of the crude product directly with methanol produced sucrose due to a deesterifi-
cation catalysed by the potassium soaps). The washing
and extraction stages resulted in about 12 to 20% loss
of product, with the "tetra" and "penta" esters, respectively, whereas the "tri"ester products were
highly miscible with methanol. The characteristics

of typical products are summarized in Table I.
Two resin types were investigated, including epoxy
esters from stepwise reaction of sucrose-partial drying
oil ester with cyclic anhydride and diepoxide, and
polyurethanes from sucrose-partial drying oil ester
and diisocyanate. Details of synthesis of the resins
and their evaluation as coatings are discussed below.

Epoxy ester-type resins from diepoxide

Synthesis. The synthesis of epoxy ester type resins from diepoxide (2) is illustrated in Figure 1.
In the first stage, a sucrose-partial drying oil ester
of a certain grade of purity (see Table I) was reacted
at 100°C for 2 hours under nitrogen in a suitable solvent with 2 mole proportion of a cyclic anhydride e.g.
phthalic, tetrahydrophthalic or dodecenyl succinic anhydrides, to give a dihalf ester intermediate, on average as indicated by acid value determination. The res-
in-forming stage involved reaction of a dihalf ester
intermediate with up to 1 mole proportion of a diepoxide, e.g. bisphenol A diglycidyl ether, its oligomers
or vinyl cyclohexene dioxide. The reaction temperature was usually below 120°C although with pentasubsti-
tuted ester, up to 140°C could be used to achieve a faster rate of reaction. The progress of resin formation
was followed by loss of carboxyl groups. Gel permea-
tion chromatographic (G.P.C.) analysis showed that resins with a range of molecular sizes including high
molecular weight fractions were present.
Resins were prepared from crude sucrose esters,
soap-free sucrose esters and the insoluble fractions
obtained after methanol extraction of soap-free sucrose
esters. Details of resins from "tetra"/"penta" esters
are given in Tables II and III. Starting from crude
esters, the potassium soaps present catalysed the reactions with carboxyl and epoxide groups. However,
with soap-free esters or methanol insoluble fractions,
a tertiary amine catalyst, e.g. about 0.25% triethylamine or benzyldimethylamine, was required to obtain a
satisfactory rate of reaction. Reaction times were
usually about 6 to 8 hours.
In many cases, the extent of reaction (from acid
value determination) was 85-90%. However, in some
cases as low as 70% conversion of carboxyl groups was
noted, e.g. in the reaction using a sucrose pentadehy-
drated castor ester containing potassium soaps (but no
tertiary amine), tetrahydrophthalic anhydride and bisphenol A diglycidyl ether with no substantial reduction in acid value after a few hours further heating.

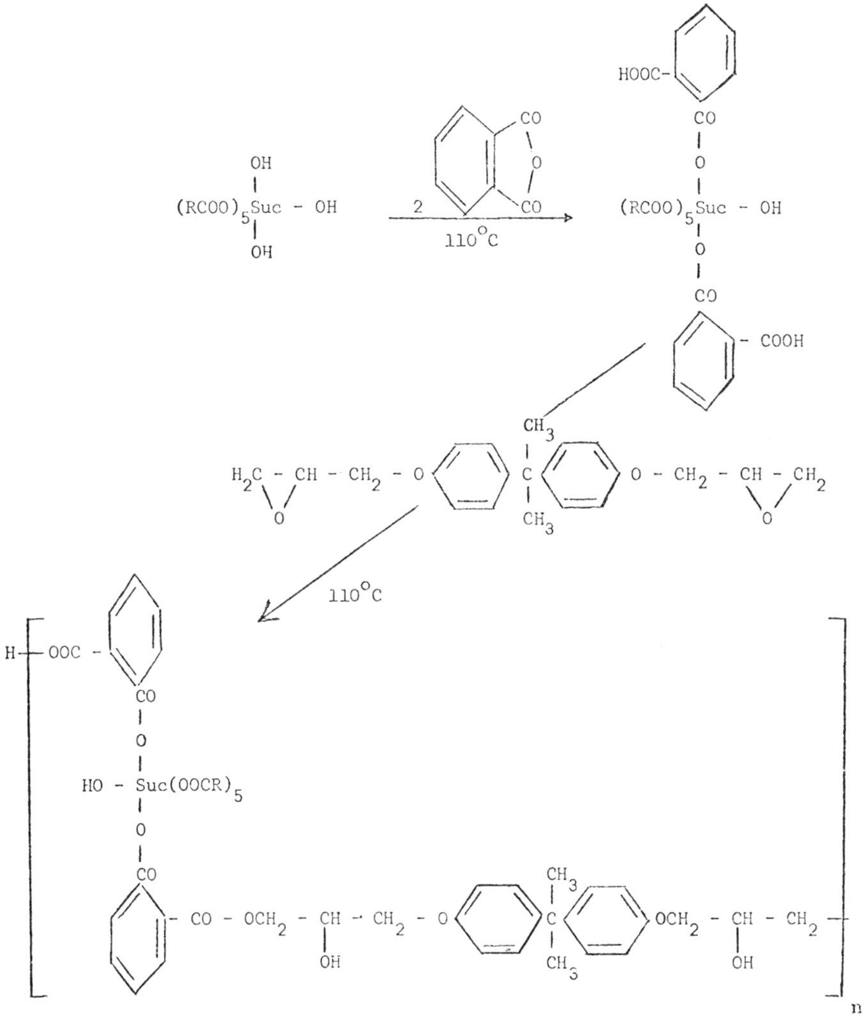

Figure 1. Sucrose diacid/diepoxide route to sucorse resins (Suc = sucrose)

TABLE I.   Characteristics of Crude and Upgraded Sucrose Drying Oil Ester Products

Sucrose ester (ratio of methyl ester/sucrose in reaction mixture)	% K₂CO₃ catalyst (on total reactants)	% K soaps in crude product (by ashing at 700°C)	% Extent of reaction (from CH₃OH evolved³)	% unreacted methyl esters (by molecular distillation)	Acid value of soap-free sucrose ester mg KOH/g	Methanol insoluble fraction from extraction of soap-free ester			
						% Yield	Acid Value mg KOH/g	Hydroxyl Value mg KOH/g	Saponification Value mg KOH/g
"Penta" ester (5:1)									
Dehydrated castor	0.5	2	c.90	11	4.1	88.5	2.1	117	165
	1.7	3.7	"	13	6.3				
	4.0	7	"		14				
3:1 Soyate/linseedate	1.0	2.7	c.90		4.7	88.5	2	115	–
Soyate	0.33	1.4	"	11	3.6	87.5	2.3	–	–
	0.5	2	"		4.3				
"Tetra" ester (4:1)									
Soyate	0.5	2	c.90	12	4.6	80	2.3	–	–
	1.0	3	"		5				
3:1 Soyate/linseedate	1.0	2.7	c.90		4.8				
3:1 Linseedate/Tungate	1.0	3	c.90		4.8				
"Tri" ester (3:1)									
Linseedate	1.0	2	c.90		5.1				
Soyate	0.3	2	c.90		4	Large proportion of product miscible with 3 volume of methanol			

TABLE II. Synthesis by Diepoxide Route – Properties of Resins Derived from "Tetra" Esters

Sucrose Ester Type	Purity	Cyclic anhydride (molar proportion)	Diepoxide (molar proportion)	Other modification (molar proportion)	Acid Value mg KOH/g (on solid Resin)	% Reaction of epoxide and COOH groups (from acid value)	% Reaction of epoxide groups (from epoxide value)	Molecular size (Å) distribution (by G.P.C.)
Soyate	Crude (3% soaps)	THPA (2)	Epikote 834 (1)	–	9	82	–	–
		THPA (1.5)	Epikote 1001 (1)	Dimer fatty acids (0.5)	12	78	–	up to 9,000 broad max. 100–400
		Dodecenyl succinic anhydride (2)	Epikote 828 (1)	–	5.5	89	–	up to 3,500 broad max. 100–400
3:1 Soyate/ Linseedate	Crude (2.7% soaps)	THPA (2)	Epikote 1001 (1)	–	15	65	–	–
		PA (2)	VCD (1)	–	12	81	–	Up to 2,000 broad max. 100–400
		PA (2)	Epikote 828 (0.75)	Butyl glycidyl ether (0.5)	10	81	92	Up to 3,000 broad max. 100–500
	Soap free (acid value 4.8)	PA (2)	Epikote 828 (0.75)	Butyl glycidyl ether (0.5)	10	–	–	–
3:1 Linseed- ate/tungate	Crude (3% soaps)	Dodecenyl succinic anhydride (2)	VCD (1)	–	10	84	–	Up to 1,500 broad max. 100–250

THPA = Tetrahydrophthalic anhydride; PA = Phthalic anhydride; VCD = Vinyl cyclohexene dioxide

TABLE III.   Synthesis by Diepoxy Route – Properties of Resins Derived from "Penta" Esters

Sucrose Ester		Cyclic anhydride (molar proportion)	Diepoxide (molar proportion)	Other modification (molar proportion)	Acid Value mg KOH/g (on solid resin)	% Reaction of epoxide and COOH groups (from acid value)	% Reaction of epoxide groups (from epoxide value)*	Molecular size (A) distribution (by G.P.C.)
Type	Purity							
Dehydrated castor	Crude (3.7% soaps)	THPA (2)	Epikote 828 (1)	–	–	69	87	Up to 20,000 broad max. 94–2,800
		THPA (2)	VCD (1.1)	–	7	87	–	Up to 6,000 broad max. 100–400
	Soap free (acid value = 4)	THPA (2)	Epikote 828 (1)	–	12	72	–	Up to 10,000 broad max. 100–400
		PA (2)	Epikote 828 (0.75)	Butyl glycidyl ether (0.5)	11	85	–	–
	Methanol insoluble from soap-free ester	THPA (2)	Epikote 828 (1)	–	12	75	–	Up to 10,000 broad max. 100–500
Soyate	Crude (2% soaps)	THPA (2)	Epikote 828 (1)	–	–	70	–	Up to 3,000 broad max. 100–400
3:1 Soyate/ Linseedate	Crude (2.7% soaps)	THPA (2)	Epikote 828 (1)	–	–	75	–	–

* By reaction with hydrochloric acid/pyridine

Epoxide value determinations on this and a"tetra"ester-
based system showed that epoxide groups were involved
in reactions other than with carboxyl groups (see
Tables II and III).
   The reaction solvent was chosen so that a homogenous reaction mixture was obtained.   This depended on
the type of diepoxide and cyclic anhydride used.
White spirit or xylene were suitable solvents with bisphenol A diglycidyl ether (Epikote 828), xylene with a
low molecular weight oligomer, Epikote 834, and a stronger solvent such as Cellosolve acetate (optionally with
xylene) with a higher molecular weight oligomer, Epikote 1001.   Xylene was required in reactions with
phthalic anhydride but white spirit could be used with
tetrahydrophthalic anhydride and dodecenyl succinic
anhydride.
   Reaction of a sucrose drying oil ester/dihalf ester derivative containing strong acid groups e.g.,
phthalic or maleic, with a diepoxide led to substantial
decomposition and charring of the sucrose moiety.
This was avoided by heating all three reactants together in a one-stage reaction.
   In reactions of strong acid groups with the simple
diepoxide, bisphenol A diglycidyl ether, evidence was
obtained of side reactions between carboxyl groups and
hydroxyl groups formed by ring-opening of the epoxide.
This produced intermediates of increased functionality
and it was found that increasing amounts of gel were
formed with proportions of diepoxide above 0.75 mole
up to about 1 mole partial sucrose ester/diacid derivative.   Gelation was more pronounced using diepoxide
oligomers which initially have at least one free hydroxyl group per mole in addition to those formed by
reaction of the epoxy group.   Gelation was avoided by
using a mixture of a diepoxide and a monoepoxide such
as butyl glycidyl ether.
   Resins based on tetrahydrophthalic and dodecenyl
succinic anhydride (but not phthalic anhydride) were
white spirit tolerant.   The solutions, however, were
rather viscous at about 50-60% concentration and, to
obtain compositions with suitable brushing properties,
it was necessary to add diacetone alcohol (or butanol)
to give a 3:1 ratio of white spirit to diacetone alcohol in the final solution.
   Evaluation of the surface coating properties of
resins derived by diepoxide route. Unpigmented coatings
containing cobalt naphthenate drier (0.05% cobalt as
naphthenate on the drying oil content of the resin)
were applied by brush or doctor blade to various substances including aluminium, glass and polyethylene

terephthalate film (Melinex) for test purposes.   The
dried coating thickness was about 25µ.   Scratch hardness was measured by the maximum resistance to scratch-
ing of coatings on aluminium with a 1 mm diameter ball
under varying loads according to British Standard 3900
Part E2.   Flexibility was assessed qualitatively by
creasing coatings on Melinex.   Resistance also was
evaluated to immersion in water and aqueous alkali at
20°C and, in certain cases, 10% acetic acid, 0.5% aqueous sulphur dioxide and 10% aqueous phosphoric acid.
Gloss was measured as specular reflectance at 45°
according to B.S. 3900: Part D2.   Resistance to cracking and detachment from aluminium substrate was measured by the cupping test,when subjected to gradual de-
formation by indentation.   Gloss changes were measured
during artificial weathering in a Marr Weatherometer
according to BS 3900: Part F3.   For an alkyd paint,
1,000 hours' exposure corresponds to about 1 1/2 years'
exposure in a temperate climate.

Various resins prepared from sucrose trisoyate or
sucrose trilinseedate, 2 mole proportions of dodecenyl
succinic anhydride, tetrahydrophthalic anhydride or
phthalic anhydride and 1 mole bisphenol A diglycidyl
ether or oligomer were evaluated.   In general, in terms
of flexibility after ageing, air-dried coatings from
these materials were of inferior quality.   Resins
from "tetra" or "penta"esters showed better performance
and the results are summarised in Tables IV and V.
Resins containing wholly drying oil substituents, e.g.
linseed and tung, became brittle and yellow on ageing,
unlike semi-drying oil types, e.g. soya and dehydrated
castor oil.

The presence of potassium soaps in resins synthesised from crude sucrose esters had an adverse effect
on the water resistance of air-dried films.   The gloss
of paints also was reduced.   Improved performance was
noted using soap-free sucrose esters or, particularly,
the products obtained by extraction of these materials
with methanol to remove the more hydrophilic mono- and
disubstituted sucrose esters, free fatty acids and unreacted methyl esters.

Resins from "penta"-substituted esters were superior to corresponding materials from lower substituted
esters.   The best coatings were obtained from resins
based on sucrose "penta"-dehydrated castor esters.   A
titanium dioxide paint of a system of this type, therefore, was evaluated in more detail. For comparison purposes,a typical drying oil alkyd paint also was tested.
The results (Table VI) show that the performance of
both resins is similar.

TABLE IV.   Synthesis by Diepoxide Route - Properties of Unpigmented "Tetra" Sucrose Ester Based Resin Films

Components of Resin (molar proportions used)				Surface Coating Properties (25 microns film thickness; 0.05% cobalt as naphthenate aged 1 week at 20°C)				
Sucrose Ester	Cyclic anhydride	Diepoxide	Other modification	Touch dry time	Scratch hardness	Water resistance	1% NaOH resistance	Flexibility
Crude soyate with 3% soaps (1)	THPA (2)	Epikote 834(1)	-	3	1100g	moderate	-	good
	THPA (1.5)	Epikote 1001(1)	Dimer fatty acids (0.5)	3	900g	moderate	-	-
	Dodecenyl succinic	Epikote 828(1)	-	3	1300g	moderate	-	good
Crude 3:1 soyate/ linseedate with 2.7% soaps (1)	THPA (2)	Epikote 1001(1)	-	$2\frac{1}{2}$	1500g	moderate	-	inferior
	PA (2)	VCD (1)	-	$2\frac{1}{2}$	1400g	moderate	-	-
	PA (2)	Epikote 828(0.75)	Butyl glycidyl ether (0.5)	$2\frac{1}{2}$	1150g	good	breakdown after $\frac{3}{4}$ hr	-
Soap free product from above (1)	PA (2)	Epikote 828(0.75)	Butyl glycidyl ether (0.5)	$2\frac{1}{2}$	1200g	good	breakdown after 1 hr	-
Crude 3:1 Linseedate/ tungate with 3% soaps (1)	Dodecenyl succinic (2)	VCD (1)	-	2	1000g	inferior	-	embrittled on ageing

TABLE V.   Synthesis by Diepoxide Route – Properties of Unpigmented "Penta" Sucrose Ester Based Resin Films

| Sucrose Ester | Components of Resin (molar proportions used) | | | Surface Coating Properties (25 microns film thickness; 0.05% cobalt as naphthenate aged 1 week at 20°C) | | | | |
	Cyclic anhydride	Diepoxide	Other modification	Touch dry time	Scratch hardness	Water resistance	1% NaOH resistance	Flexibility
Crude dehydrated castor with 3.7% soaps (1)	THPA (2)	Epikote 828(1)	–	2	1600g	good	breakdown after 2 hrs.	good
	THPA (2)	VCD (1.1)	–	2	1700g	good	inferior	good
Soap-free product from above (1)	THPA (2)	Epikote 828(1)	–	2	1600g	very good	breakdown after 3½ hrs.	good
	PA (2)	Epikote 828(0.75)	Butyl glycidyl ether (0.5)	2	1500g	very good	breakdown after 3½ hrs.	good
Methanol insoluble from soap-free ester (1)	THPA (2)	Epikote 828(1)	–	2	1600g	very good	breakdown after 6 hrs.	good
Crude soyate with 2% soaps (1)	THPA (2)	Epikote 828(1)	–	2	1300g	very good	breakdown after 1½ hrs.	good
Crude 3:1 soyate/ linseedate with 2.7% soaps (1)	THPA (2)	Epikote 828(1)	–	2	1400g	very good	breakdown after 1½ hrs.	inferior

TABLE **VI**. Evaluation of Titanium Dioxide Paints based on Resin from
Sucrose'Penta' - D.C.O. Esters (3% K Soaps), Tetrahydrophthalic
Anhydride and Bisphenol A Diglycidyl Ether and, for Comparison,
Corresponding Paint from Commercial D.C.O. Alkyd Resin

0.55:1 Pigment/binder: Air-dried 1 week at $20^{\circ}C$
0.05% Co as naphthenate : 25 microns film thickness

T e s t	Paint from Sucrose Resin	Paint from Commercial Alkyd
Drying time (finger touch)	$1\frac{1}{2}$ to 2 hours	2 to 3 hours
Scratch hardness (on aluminium)	just passed 1600g	just passed 1700g
Flexibility (on Melinex) after 1 week after 1 year	very good " "	very good " "
Water:immersion on Melinex at $20^{\circ}C$/10 days	only slight softening and dulling	as opposite
1% aq NaOH:immersion on Melinex at $20^{\circ}C$	incipient breakdown after 3 hours	incipient breakdown after $1\frac{1}{2}$ hours
0.5% aq SO_2:immersion on Melinex at $20^{\circ}C$/4 days	Microblistering and softening but no dulling	dulled and softened with loss of adhesion
10% H_3PO_4:immersion on Melinex at $20^{\circ}C$/4 days	Slight softening and dulling	as opposite
10% Acetic acid:immersion on Melinex at $20^{\circ}C$/1 week	No apparent effect	as opposite
Cupping test (on aluminium)	Paint failed at 5mm depth of indentation	Paint failed at 7mm depth of indentation
Gloss retention (45° specular reflectance)		
Initially	90	95
After 1000 hours artificial weathering	40	55
After 1 year natural weathering (Teddington)	50	60

Polyurethane Resins

Sucrose polyurethane oils produced by direct reaction of a diisocyanate with a sucrose partial ester,
having an average degree of substitution of 4.5 to 6,
were studied briefly by Bobalek, et al (3) several
years ago. The results were rather inconclusive. For
example, air-drying polyurethanes were obtained with
relatively poor drying properties, the coatings were
brittle and yellowed excessively. Difficulty was experienced in reaction of the stoichiometric amounts of
diisocyanate required for optimum film performance,
without gelation.
     In the present work, it was found that about 0.85
mole toluene diisocyanate was the maximum which could
be reacted with a sucrose "pentaester"without gelation.
At this level of addition, the product showed reasonable resistance to aqueous alkali, apparently better
than the materials reported above but inferior to
commercial products. For example, air-dried polyurethane oil coatings derived from the methanol-insol-
uble fraction of soap-free sucrose pentasoyate or pen-
tadehydrated castor esters lost 15% and 5% by weight
respectively on immersion for 2 hours in 5% NaOH. Resins with considerably better performance were achieved
by blocking, on average, one hydroxyl group of the
sucrose pentaester prior to reaction with diisocyanate.
Various approaches were investigated including reaction
with: (1) phenyl isocyanate; (2) cyclic anydride and
then monoepoxide with the half ester formed; (3) 1:1
chelate complex derived from aluminium isopropoxide and
acetyl acetone; and (4) cyclic anhydride and then ben-
zylation of the half ester.
     Methods 1, 2 and 3 are simple, essentially one-pot
reactions and no work-up of the product is required.
Method 4 however, is unattractive since a washing
stage is necessary to remove potassium chloride formed
by reaction of benzyl chloride and the potassium salt
of a sucrose ester/half ester derivative of, e.g.tetra-
hydrophthalic or maleic anhydride.
     The maximum amount of toluene diisocyanate which
could react without gelation varied according to
the type of modification used e.g., up to 1.35 mole
proportions with the mono-/diisocyanate route and somewhat lower (1.1 to 1.2 mole) proportions in the other
cases. The mono-/diisocyanate approach, as expected,
produced resins with the best alkali resistance, in
some cases at least comparable to high quality, commercial, air-drying polyurethane resins and, therefore,
was studied in more detail.

Synthesis by the monodiisocyanate route. The reaction was carried out in two stages in the same vessel using soap-free "penta"-substituted soya, dehydrated castor and 3:1 soya/linseed esters, and "tetra"-
substituted soya ester or, in certain cases, the insoluble fractions from methanol extraction. The sucrose
ester was dissolved in xylene and heated with 1 molar
equivalent (or 2 equivalents with the "tetra" ester) of
phenyl isocyanate at 90-100°C and dibutyl tin dilaurate
as catalyst. On a few gram scale, the reaction was
complete in 2-3 hours as indicated by the absence of
NCO peak at about 2260_{cm}^{-1} in the infrared spectrum.
With larger (greater than 25g) scale preparation, addition of a larger amount of catalyst, up to 1% by weight
of reactants, was required to achieve a satisfactory
reaction time of about 4-5 hours. Also, to avoid separation of intractable solid, phenyl isocyanate was
added in portions. In the final stage of adding diisocyanate, preliminary small scale tests were carried
out to determine the maximum quantity that could react
at 95-100°C without gelation.

Soaking tests were carried out on unpigmented film
containing 0.05% cobalt as cobalt naphthenate to determine the resistance of resins from various sucrose esters to immersion in 5% aqueous alkali. Air-dried
(aged for 2 weeks at 20°C) and stoved(120°C/30 minutes)
films on Melinex were examined. The former dried to a
touch-dry state in about 2 hours and after 2 weeks were
hard (scratch hardness 1400g), flexible and showed good
water resistance. Degradation was determined quantitatively from the progressive percentage loss in weight
of the solvent-free film. The results of alkali resistance tests are given in Table VII and for comparison,
data for commercial, drying oil modified, polyurethane resins. These show clearly that resins from
a sucrose "penta" ester had better alkali resistance
than those from the corresponding "tetra"-derivatives.
Those from the insoluble fractions from methanol
extraction were better than those from the corresponding soap-free esters. Resins from sucrose "penta"-
dehydrated castor esters were the best in this respect
and at least as good as commercial resins. Titanium
dioxide paints of this system were evaluated in more
detail against paints prepared from 4 commercial resins. A pigment/binder ratio of 0.55:1 was used with
0.05% cobalt as naphthenate and methyl ethyl ketoxime
as anti-skinning agent. For test purposes, paints
were applied to Melinex, chromated aluminium and phos-
phated steel. The results (see Tables VIII, IX and
X) showed that the sucrose-based paint selected for

TABLE VII.   Alkali Resistance of Unpigmented Sucrose Polyurethane Resin Films Derived from Sucrose Esters, Phenyl Isocyanate and Toluene Diisocyanate and Comparison with Commercial Resins

Sucrose ester starting material (or commercial resin)	Purity of sucrose ester	Maximum amount of TDI without gelation	Immersion in 5% aq NaOH			
			Air-dried film		Stoved film	
			Weight loss %	hours	Weight loss %	hours
Penta soyate	(a) Soap-free	1.35	1 / 28	4½ / 21	5 / 20	18 / 111
	(b) CH$_3$OH-insoluble fraction of (a)	1.35	1 / 17	4½ / 21	8	111
Penta dehydrated Castor	CH$_3$OH insoluble fraction of soap free ester	1.1	1.7	47 / 121	2.2 / 2.9	93 / 118
Penta 3:1 soyate/linseedate	CH$_3$OH insoluble fraction of soap free ester	1.3	4.3 / 23	23 / 117	2 / 6	63 / 135
		1.1 (less than max.)	6 / 19.7	5 / 29	–	–
Tetra soyate (using 2 moles phenyl isocyanate)	Soap-free	1.4	–	–	10.5 / 67	5 / 24
Commercial soya-urethane oil		–	35 / 70	20 / 96	11 / 19	48 / 72
Commercial dehydrated castor urethane oil		–	36 / 64	20 / 90	7 / 11	72 / 96
Commercial soya urethane alkyd		–			10	48

TABLE VIII. Evaluation of Changes in Gloss Reflectance (60°C) During Weathering of Air-drying Titanium Dioxide Paints from Sucrose Polyurethane Resin (Mono/Diisocyanate Route) and Corresponding Paints from Commercial Resins

Resin used in paint	Initial gloss	Gloss after artificial weathering* (B.S. 3900 Part F3) Hours				Gloss after natural weathering – Teddington*
		500	1000	1500	2500	1 year
Sucrose penta DCO ester polyurethane	95	85	75	68	50	58
Soya urethane oil	91	72	60	55	severe chalking	32
Dehydrated castor urethane oil	96	70	59	52	"	43
Soya urethane alkyd	91	82	70	65	"	46
Drying oil alkyd	95	78	59	52	35	61

*Panels washed before test

TABLE IX.   Evaluation of Humidity, Water, Alkali and Salt Spray Resistance of Air-drying Titanium Dioxide Paints from Sucrose Polyurethane Resins (Mono/Diisocyanate Route) and Corresponding Paints from Commercial Resins

Resin used in paint	Humidity Test (200 hours) (BS 3900 : Part F2) (% loss in gloss; appearance)		Salt Spray Test (400 hours) (BS 3900 : Part 4) (% loss in gloss; appearance)		Water resistance (100°C/5½ hours) (on Melinex)	Alkali resistance (5% aq.NaOH) (on Melinex)
	Aluminium	Steel	Aluminium	Steel		
Sucrose penta DCO ester polyurethane	6.5 no surface defects	5.7; microblistering over whole panel	3.3; no surface defects	0.0; isolated microblistering	Film yellowed; no surface defects	under 10% weight loss after 122 hours
Soya polyurethane oil	5.2; isolated micro-blistering	60; microblistering over whole panel	9.4; no surface defects	0.0; microblistering along edges of scratch	Film yellowed; no surface defects	About 10% weight loss after 30 hours; film detached on washing
Dehydrated castor polyurethane oil	41.2; micro-blistering along edges of scratch	42.2; microblistering over whole panel	6.2; no surface defects	0.0; microblisters along edges of scratch	Film yellowed; no surface defects	result as immediately above
Soya urethane alkyd	43.4; micro-blistering over whole panel	60.2; microblistering over whole panel	11.1; no surface defects	0.0; microblisters along edges of scratch	Film yellowed; no surface defects	result as immediately above
Linseed epoxy ester	6.3; no surface defects	32.2; microblisters over whole panel	6.3; isolated microblistering	1.1; isolated microblistering	Film yellowed; no surface defects	About 8% weight loss after 122 hours

TABLE X.   Evaluation of Mechanical Properties of Air-drying Titanium Dioxide Paints from Sucrose Polyurethane Resin (Mono/Diisocyanate Route) and Corresponding Paints from Commercial Resin

Resin used in paint	Adhesion by Cross-hatch test (B.S. AU 148)		Taber Abrasion Loss (mg) after 1000 cycles (U.S. Standard No.141a method No. 6192)	Cupping Test Depth of indentation (mm) (B.S. 3900 Part E4)		Elongation Test Minimum % elongation that produces cracks (ASTM D522-60)
	Aluminium	Steel		Aluminium	Steel	
Sucrose penta DCO ester polyurethane	Inferior	Inferior	67.5	4.2	3.7	6.3
Soya urethane oil	Fair	Fair	Failed* at 300 cycles	6.0	7.9	no cracks produced
Dehydrated castor urethane oil	Fair	Good	Failed* at 600 cycles	6.4	>12.0	no cracks produced
Soya urethane alkyd	Fair	Good	87.6	6.7	10.3	no cracks produced
Linseed epoxy ester	Fair	Fair	65	6.5	11.6	no cracks produced

* Coating worn down to substrate

test, had similar or better chemical resistance properties and weathering performance, but lower adhesion and
flexibility than the commercial polyurethane oils and
linseed epoxyester paints.
    Synthesis from sucrose ester/half-ester deriva-
tives, monoepoxide and diisocyanate. Sucrose "penta"-
soyate (the methanol insoluble fraction of the soap-
free product) was reacted at 100°C for 4 hours with a
mixture in xylene of 1 mole equivalent cyclic anhydride
e.g., phthalic or tetrahydrophthalic anhydride
and monoepoxide e.g., styrene oxide or Cardura E (the
glycidyl ester of versatic acid), branched C9-Cfatty
acids, in the presence of N-benzyl dimethylamine as
catalyst.   A faster rate was achieved with the phthalic system.   The course of reaction between half ester
carboxyl groups and epoxide groups was followed by acid
value.   About 90% reaction was achieved in all cases
and to the product was added 1.0 mole toluene diisocyanate (the maximum amount without gelation) and dibutyl tin dilaurate as catalyst.
    Air-dried or stoved (120°C/30 min) coatings con-
taining cobalt as naphthenate showed alkali resistance
comparable to that of resins from soap-free soya esters
derived by the mono-/diisocyanate route(see Table VII).
    Synthesis from sucrose ester, aluminium diiso-
propoxide monoacetylacetonate and diisocyanate. In preliminary experiments, sucrose "penta"soyate or sucrose
"penta" dehydrated castor ester reacted in toluene at
110°C with an equimolar proportion of aluminium diisopropoxide monoacetylacetonate. (The latter was prepared separately by reaction of equimolar proportions of
aluminium isopropoxide and acetylacetone in refluxing
toluene with elimination of 1 mole 2-propanol).   The
modified sucrose ester reacted with 1.1 mole toluene
diisocyanate (the maximum without gelation) and dibutyl tin dilaurate as catalyst.   The performance of
air-dried or stoved coatings was similar to material
derived by the mono-/diisocyanate route in terms of
alkali resistance, but they were more brittle.
    In a modification of this approach, tractable res-
ins also were produced using aluminium isopropoxide
for blocking sucrose hydroxyl groups, prior to reaction
with toluene diisocyanate.   Coatings from this resin
also tended to be brittle.
    The diacetylacetone derivative of aluminium iso-
propoxide failed to react with the sucrose drying oil
ester, presumably due to steric factors.

Discussion
    The present work has demonstrated the feasibility

of making good resins from sucrose higher drying oil
esters by relatively simple reactions which can be carried out in conventional resin-making equipment.   With
the sucrose epoxyester system, one of the best resins
was sampled to several members of Paint Research Association for evaluation.   Their assessment confirmed
that performance was equivalent to high quality alkyds
but the price was too high.   Improved performance, e.g.
alkali resistance, was obtained using diepoxide oligomer but, because of its higher equivalent weight, led
to more expensive resins.   In addition, they were
brittle and "strong" solvents were required.

The second class of resin investigated, sucrose
drying oil ester-polyurethanes in which the expensive
bisphenol A diepoxide is replaced by, for example,
toluene diisocyanate at about half the cost per ton,
are commercially more attractive.   They are cheaper,
have better performance, are more compatible with
white spirit and with air-drying systems and avoid the
need for "strong" solvents which could lead to difficulties in applying top coats.   The best resin, based
on sucrose "penta"-dehydrated castor oil esters with
17% combined sucrose and 14% combined toluene diisocy-
anate/phenyl isocyanate, was at least comparable to a
conventional, air-drying polyurethane alkyd.   A lot of
development work, however, is still required to provide
paints with optimum properties in terms of gloss, gloss
retention and durability, drying time, color, flow,
brushability, pigment dispersion properties, etc.

Resins obtained using alternative means to phenyl
isocyanate for blocking the hydroxyl functionality of
a sucrose "penta" ester, e.g. reaction with aluminium
diisopropoxide monoacetylacetonate or cyclic anhydride/
epoxide also show promise.
There is considerable support for the view that
organic solvent soluble, drying oil modified resins represent a good market,certainly during the next decade,
and are more likely to achieve exploitation under present economic conditions.

Another area for investigation, which has not yet
been explored in detail, is the use of sucrose drying
oil esters in two-pack or moisture-curing polyurethanes.
These types should provide the superior abrasion and
chemical resistance required for high performance application,e.g. coatings for chemical plants,floors, etc.

To encourage the paint and resin manufacturers to
take a serious interest in sucrose drying oil esters,
the immediate priority is to produce them by a practical process which does not use expensive solvents.
Preliminary studies towards this objective have demon-

strated the feasibility of making sucrose higher drying oil esters directly from sucrose and methyl esters
in absence of solvent. Products, as methanol insoluble fractions, have been obtained in 75-80% yield and
efforts are now being made to overcome the problems
of dark color and long reaction times. We hope that
success in this project will persuade the paint and
resin industry that sucrose drying oil esters have a
bright future.

Acknowledgements

This paper is based on work carried out under contract for the International Sugar Research Foundation,
Inc. The assistance of Dr. R. Iyer, Dr. N.R. White-
house and other colleagues in the Paint Research Association and the permission of the Council and Director
to present this paper is gratefully acknowledged.

Abstract

In seeking new opportunities for sucrose in the
paint industry, the feasibility of making practical sucrose resins has been demonstrated. Studies have been
concerned mainly with air-drying systems derived from
sucrose partial drying oil esters containing linseed,
soya or dehydrated castor oil unsaturation. In one
approach, resins were obtained by step-wise reaction of
these materials with a cyclic anhydride e.g. tetrahydrophthalic or phthalic anhydride, to give a di-half
ester intermediate, and a diepoxide, e.g. bisphenol A
diglycidyl ether. Titanium dioxide paints from the
best resins give performance comparable to conventional
drying oil alkyd paints with respect to gloss retention
during artificial and outdoor weathering tests, water
and alkali resistance during immersion, scratch resistance and flexibility. In more recent work, another class of air-drying resin has been made which is
cheaper and has better chemical resistance. For example, the weight loss of films immersed for 120 hours
in 5% aqueous potassium hydroxide was under 8%. In
general, the best resins were obtained from tetra/
penta esters with conjugated unsaturation which were
upgraded by washing with dilute acid to remove soaps
and extraction with methanol to remove lower esters.

References

1.   Kollonitsch, V., Sucrose Chemicals, The Interna-
     tional Sugar Research Foundation, Inc., Bethes-

da, Maryland, (1970) pp 52, 192.
2. Faulkner, R. N., (to Research Corporation),
British Patent 1,424,862 equivalent to U.S.
Patent 3,870,664, (1975).
3. Bobalek, E. G., De Mendoza, A. P., Causa, A. G.,
Collings, W. J., and Kapo, G., Ind. Eng. Chem.,
Prod. Res. & Dev., (1963), 2 (1), 9-16.

Biographic Notes

Raymond N. Faulkner, B.Sc., F.R.I.C., Deputy Head
of Chem. Res. Educated at Durham Univ. Joined the
Paint Research Association in 1948, studying surface
coatings and photopolimerization. The Paint Research
Association, Waldegrave Road, Teddington, Middlesex
TW11 8LD, England

14

Prospects for Sugar in Surface Coatings

JOHN C. WEAVER

Consultant, 3305 Enderby Rd., Shaker Heights, Ohio 44120

Using sugar in paint would have seemed unlikely if not downright foolish to Thomas Childs when in about 1737 he brought the "Boston Stones" over from England to begin paint manufacture in America. He or his apprentice rolled a two foot stone ball back and forth to disperse pigment in oil in a stone trough. These stones, off Union Street, now mark the hub of Boston and are symbols of the modern Federation of Societies for Coatings Technology. But, paint science and technology have changed immensely, and it is time for new perspectives toward opportunities for sucrose in paints.

The historic recipe for durability in paints required only water repellant media such as the pitch on Noah's Ark and naval stores treatment of wooden ships through the successive millenia thereafter. Linseed oil had a long and honorable record in both artistic and protective paints, as well as in linoleum and in those yellow slickers which protected New England fishermen against northeasters. The Chinese have a venerable history of using oil from tung nuts in waterproofing their boats. Of the three major vegetable products, fats, carbohydrates and proteins, only this class of fatty oils had major use in paints, while carbohydrates were too intractable and proteins were too vulnerable for all but minor use in whitewash or decor-

ative distemper paints.   Intractability implies a durability of sorts in the ether linkages of cellulose
which survive centuries of exposure in superior wood
surfaces at very low erosion rates, albeit of drab appearance.
   One major exception, while minor to the immense
bulk of the big three, is the broad class of natural
resins and gums.   The wide range from fossil amber to
contemporary rosin covers a complexity of terpene chemistry beginning with simple isoprene and extending
through terpenes such as the pinenes and dipentene to
multi-ring,oxidized structures rigid enough to enhance
the weak flexibility of a paint based only on an oil.
Kauri gum from New Zealand was a standard of excellence
among these natural resins for many decades, until the
beginnings of modern polymer chemistry displaced these
chance mixtures of nature.   Leo Baekland's phenol
formaldehyde resins were the first major effort toward
designed polymers to start the synthetic plastics industry.   Side chain substitutions of these phenolics
made them soluble in drying oils, and these combinations, along with plasticized nitrocellulose, started
the paint industry along synthetic routes.   Only a few
years later the first alkyds appeared commercially, and
over the last fifty years these alkyds largely have
displaced the old oil and resin combinations.
   Latex, synthesized in Germany in World War II in
several formula variations,was substituted for unavailable natural latex in rubbers and also for drying oils
and resins in paints.   These, plus designed polymer
evolutions from the American Rubber Reserve program of
several chemical species of latexes, educated the
householder to the convenience of water system paints
which were far better than earlier, lime-based, water
dispersed paints.
   Air pollution and political reaction, first in the
Los Angeles basin then in the San Francisco Bay area,
started a series of regulations not only on automobile
and factory exhausts, but also on most kinds of sol-
vent-based paints.   Quickly the technologies of latex
paint and occasional, water reducible alkyds from the
1940's were tried for all manner of paint applications,
both trade sales and industrial.   While both kinds
were successful then, for interior flat wall paints,
neither served well in exterior, high gloss paints and
their success in these, more difficult uses only now is
emerging in the 1970's.
   Sucrose utilization research has been directed
mainly toward use in place of glycerol and other polyols in alkyds for solvent systems.   This was in keep-

ing with the times in paint technology and can continue
to be a viable opportunity. It is hoped that the ongoing research stemming from the works of Hass,Stewart,
Herstein, Bobalek and Faulkner, among others, soon will
reach commercial fruition. The Arabian oil crisis in
1973 seems to have made lasting the much higher prices
of glycerol, pentaerythritol and other polyols, and
improved the competitive chances of that purest, driest
polyol, called sucrose.
     The new world of paint technology requires, how-
ever, a reassessment of the opportunities of sucrose,
in respect to several major trends in pollution and
energy control, consumer preference for water-based
paints and advances in polymer science and technology.
These trends include;

     Water dispersible alkyds and polyesters
     Water dispersible urethanes and other prepolymers
     High "solids" coatings, 80% minimum
     Reactive oligomers
     Sophistication of polymer architecture
     Specialization in industrial end uses

     We can assess these better after a brief review of
general trends in the industry, and how supportive, industrial research is pursued.

The Coatings Industry in Transition

     The sugar industry's business and chemical person-
nel need perspectives from various heights to plan better their general strategy and specific R & D programs
toward using sucrose in coatings, among other non-edi-
ble products. The worlds of sugar and paint are different, of course, but some general comparison of their
magnitudes is useful, as in Table I.

Table I.   Comparison of Sugar vs. Paint Production
            (Millions of Metric tons/year)

	Sugar	Paint	Paint Polymers
North America	16.8	5	2.5
Europe, Western	12.2	4	2
Europe, Eastern	12.1	3	1.5
Asia, Except China	12.7	2	1
Other	25.3	1	.5
Total	79.1	15	7.5

A tonnage-minded, sugar entrepreneur could scorn
paint just as the petroleum people did in their first
half century. But, by 1930, most petroleum refiners
had excess fractions of gases, light distillates and
still bottoms which reduced their profits and necessi-
tated their invention or innovation of new products.The
lighter fractions were worked over to make a diverse
range of oxygenated solvents which made possible the
formulation of much better lacquers for automobiles and
furniture.   The still bottoms, under such intriguing
names as CTLA polymer (For clay tower, Louisiana), were
offered wistfully to the paint trade as equal to linseed oil for drying.   They most emphatically were not,
though certain much refined versions of them still
find minor uses in aluminium paints, etc.

In 1943, while World War II put great pressure on
the Rubber Reserve organization complex to produce synthetic rubber fast and well, an Exxon chemist, the late
Dr. William Sparks, threatened to take an afternoon off
and design polymers specifically for paints from simple, pure butadiene and related comonomers.   Many
years and millions of dollars later he succeeded, but
only in narrow, specialized coatings such as flame cur-
ed can liners.   The financial return of this one venture was not huge, but it sparked other ventures along
similar lines, i.e., to start with pure monomers to design polymers for very specific end uses, hopefully
large and profitable.   That is my main theme here.
The Achilles Heel of that venture was mainly residual
olefinic unsaturation, i.e., continuing chemical reaction and film degradation.   It could not compete with
the classical alkyds, with all their versatility across
a wide range of paint uses.

Paint science and technology have become very sophisticated and specialized, as well as competitive,
and rely more on proprietary technical superiority than
on patent protection, even though each is important.

Marketing paint is far different from marketing
sugar and some insight on paint R & D ·methods is important to use of sucrose therein.   A paint company
does not have to be huge to be successful.   One fine
example is Midland Industrial Finishes, a division of
Dexter Corporation, with a world-wide reputation for
excellence in silicone-containing polymers in industrial baking finishes.   Its technical vice-president,
Dr. Milton Glaser, is a pioneer in this field and has
earned a wide reputation for his skills in technology
and management.

Champions Are Needed

Innovation through the use of champions and good
communications is the subject of Glaser's tutorial message (1) to his industry, even to his competitors. His
concepts, with their listings and numerical ratings,
are worthy of study by any R & D manager, especially
those rising to the challenge of getting sucrose into
paints. While innovation often stems from the mind of
just one individual, it has little chance of relevance
and success unless it fits well into a complex environment of many parts. Glaser listed in this environment
seventeen major factors (Table II), of which "New Technology" is only one.

Table II.   Environmental Factors Affecting the R & D
            Process

Macro Factors	Technical Factors
Competition	Current Technology
Economics	New technology
Survival	Patent situation
	State of the art

Organizational Factors	Individual Factors
"Climate" for R & D	Experience
Communication Processes	Intuition
Decision processes	Personality traits
Size	Risk Propensity
Sophistication of R & D	Status
	Technical competence

A good research success and its report is not
enough. Like the "commencement" ceremony upon earning
a baccalaureate, it is only a start. The innovative
researcher is part of a management team and cannot do
the whole job of innovation alone. Glaser set up some
numerical weighting factors for some of these salient
factors, and tested them against practical experience
in coatings projects (Table III).

Table III.   Most Significant Factors for Innovation

	Weighting
Effectiveness of communications(TM,TC,TT)	20
10 5 5	
Scientific and technological competence	20
Presence of a "Champion"	15
Recognition of market opportunities	15
Recognition of technical opportunities	10
Degree of top management interest	10
Competitive factors	5
Timing	5
	100

Of particular note are the high weights he gives to:

1) Effectiveness of communications

    Technical - to - Marketing   (TM)
    Technical - to - Customer   (TC)
    Technical - to - Technical   (TT)

2) Champion, defined not as a star like Babe Ruth or Olga Korbut, but rather as a person outside the research group, but very much inside the organization, with enough stature to guide, defend and **promote a project at all levels.**

70 Is a passing grade and Glaser illustrates his ratings in Table IV.

Table IV.   Heat-Resistant Silicone Coating. Assessment of Significant Factors Influencing Successful Innovation

	Weighting
Communications (TM, TC, TT)	16
9 2 5	
S/T competence (80 x 0.20 = 16)	16
Champion	11
Market opportunities	11
Technical opportunities	11
Top management interest	8
Competitive factors	2
Timing	4
Innovation Potential (IP)	79

Glaser's views on innovation are offered here,
though briefly, to show that, while coatings are com-
plicated, innovations in them occur often and a team
system is necessary to their success.   His scheme
could be extended backward a lot more.   He assumed
availability of existing polymers for his coatings,
whereas I see extra, major factors of scale up and processing costs for large penetration of the coatings industry by any sucrose oligomer or polymer which may be
chosen for exploitation.
     Innovations in coatings are happening even faster
in the 70's, as this industry and its many diverse cus-
tomers face up to higher petrochemical costs and to
stiffer environmental controls.   All the while, coatings continue to compete with other surfaces; metallic,
plastic and elastomeric.
     Sucrose derivative candidates for coatings need a
strong rationale for consideration by a coatings formulator.   He always is swamped with work and overwhelmed
by vendors' offerings of this or that polymer or additive as an improved variant of something old and familiar.   He is more likely to try one of these than something innovatively new to solve his problem, and without undue liability to his customer.   It is no discred-
it to monomeric sucrose ester plasticizers, as established components of coatings, to note that they were
close enough to earlier plasticizers to avoid undue
alarm in a coatings formulator over gross liability
from his customer, and thereby win his interest.
     Sucrose derivatives which are more radical, inno-
vative candidates in coatings will need careful selection of coatings end-use targets in order to be feasible, e.g., to get a passing grade in Innovation Potential by Glaser's rating system.   To select such targets requires both broad overview and some specific in-
sights into coatings markets.
     Coatings dollars and gallons are reported in semi-
detailed classifications by government and private organizations.   A few selections from these data are
useful to indicate trends and where innovative sucrose
derivatives might be worth trials.   United States data
are more available and represent sufficiently well its
one third or more of the world coatings market.   These
quotations are from a 1976 Stanford Research Institute
Report to the National Paint and Coatings Association
(2).   Pounds rather than metric units are still common
in the language.   The overall consumption of raw materials in recent years is shown in Table V, from the
"NPCA Data Bank Program - 1976".

Table V.   The Raw Materials of U.S. Paint
            in Billions of lb.

	1973	1974	1975
Resins	2.27	2.24	2.06
Pigments	2.64	2.49	2.31
Solvents	4.75	4.32	3.92
Additives*	0.05	0.06	0.05
Total	9.71	9.11	8.34

(*) Includes only selected additives

The National Paint and Coatings Association, 1500
Rhode Island Avenue, Washington, D.C. 20005, contracted
for its own members the preparation of this 220 page,
1976, report, recognizing that the normally, very complex coatings industry became even more so in 1973.
Price escalations on Middle East petroleum triggered
abnormal accumulations of inventories of materials for
paint, distorting the year to year trends.   Great
price increases in petrochemicals, e.g. ethylene at
3¢/lb in the 1960's, 12¢ in 1976 and projected to 20¢
in 1980, are causing producers of paint polymers and
additives to reassess their strategies.   Trends such as
this, coupled with mounting environmental restrictions
on solvents require a reassessment of paint uses of
natural products, and notably sucrose.
    Paint components are needed in hundreds of species
and subspecies to make thousands of paint formulations
in full ranges of colors to cover a great diversity of
surfaces from ship bottoms to space vehicles and a
myriad of homes, automobiles and most everything in
between.   Although about a dozen companies make about
half of all U.S. paint,there are about 1500 paint mak-
ers to be accounted for in a systematic survey, annually by the U.S. Department of Commerce,Bureau of Cen-
sus (CIR), (S.I.C. 2851), and in more accurate detail
by the Stanford Research Report (SRI) as quoted here-
after.

Table VI.   Production of Paints and Coatings
             (Millions of U.S. Gallons)

	1973	1974	1975
Trade Sales			
SRI	472	471	435
CIR	424	475	451
Industrial			
SRI	470	460	418
CIR	473	457	438
Totals			
SRI	942	931	853
CIR	897	932	889

Note: CIR = Current Industrial Reports,
      M28F, U.S. Department of Commerce,
      Bureau of the Census

The trade sales half of paint production has been
further classified by Stanford Research Institute, as
in Table VII.

Table VII.   Estimated Production of Trade Sales Paints
             and Coatings (Millions of US Gallons)

Market Segment	1973		1974		1975	
Interior, water-based		150		157		150
Flat	122		121		117	
Semi-gloss & gloss	23		30		28	
Other	5		6		5	
Interior, solvent-based		65		60		56
Flat	14		13		12	
Semi-gloss & gloss	26		24		22	
Varnish	7		6		5	
Other	18		17		17	
Exterior, water-based		98		100		87
Flat*	89		90		79	
Trim	5		6		5	
Other	4		4		3	
Exterior, solvent-based		79		**71**		62
Flat*	35		30		24	
Enamel	20		18		16	
Primer	9		8		7	
Varnish	5		5		5	
Other	10		10		10	
Miscellaneous		80		83		80
Auto & machinery refinishing						
Enamel	13		14		15	
Lacquer	8		9		10	
Primer	11		10		10	
Traffic Paints	48		50		45	
Totals		472		471		435

* Includes house paints, combination house trim paints, masonry
  finishes, and pigmented stains.

Source: SRI estimates.

Note well that the long, steady trend toward water-based paints continues unabated.   In the four
principle classes, sixty percent of water-based paints
have vehicles which mainly are latexes synthesized
from vinyl acetate and several acrylate esters, all
derived from petrochemicals such as ethylene.   The
other 40% are still made mainly from conventional sol-
vent-based alkyds, but there is new interest in trans-
forming these alkyds into water-based alkyds, where
sucrose may find new opportunities.

Water dispersible alkyds, as contrasted to solvent soluble alkyds, are likely to have higher free carboxyl and hydroxyl groups. Often they have ether groups from polyglycols. The carboxyl groups are partly neutralized, nominally by ammonia but, with more sophistication, by various amine alcohols which aid in dispersion and later in coalescence to higher gloss films. There is much activity in this field both within paint companies, and among independent resin vendors. It is timely for sugar interests to evolve a strategy to find how sucrose best can be built into these water dispersible alkyds in competition with commonly used polyols and ether glycols. Such strategy needs the full management concept of, (a) specific, paint-end use goals, (b) a systematic alkyd R & D program, (c) basic understanding of the two-phase chemistry involved (typified by such technologic terms as "HLB" for hydrophilic-lyophilic balance and the "solubility parameter concept" which is one of the newer tools of paint polymer chemists), and (d) the "champion" concept, as advocated by Glaser to carry the whole project forward.

Surfactants, thickeners and other "additives" are important minor components in all manner of paints and, in just the trade sales half of the business, are a rather large market. Those sugar interests with a stake in sucrose-based surfactants could look at opportunities in paints, particularly water-based paints, but with some caution about plunging prematurely into this highly sophisticated and competitive market. The surface chemistry and rheology relationships in pigmented coatings are too complex for any but highly experienced paint technologists.

Industrial Coatings

"Chemical Coatings" is the self aggrandizing term now used commonly for industrial paints and coatings, and implies new opportunities for sucrose in forming new reactive oligomers in one or another of the broad classes of these coatings. The SRI data base (Table VIII) lists 22 general classes within this half of the coatings market, and significant trends among them.

Industrial coatings are based on polymers much more diverse than those used in trade paints. Their trend toward water-based polymers is much slower because of a lower emotional factor and a higher product liability from hasty change. Nevertheless, the mounting prices of petrochemicals and of methane and alternate energy costs of curing films rapidly in a finishing line do imply selective trends toward water-

Table VIII.   Estimated Production of Industrial Paints and
              Coatings   (Millions of US Gallons)

Market Segment	1973	1974	1975
Wood furniture and fixtures	60	51	46
Wood and composition flat stock	27	35	28
Metal furniture and fixtures	30	29	24
Containers and closures	48	48	44
Sheet, strip, and coil	22	23	15
Major appliances	12	11	9
Other appliances	6	6	5
Automotive, OEM - topcoat	23	17	16
- primer	12	9	8
- after market + miscellaneous	10	9	8
Trucks and buses, OEM	12	10	10
Railroad, OEM	4	4	2
Other transportation, OEM	4	4	5
Machinery and equipment	30	26	24
Electrical insulation	13	13	10
Paper, film, and foil	7	7	7
Marine - pleasure, OEM	1	1	1
- new and commercial	5	5	5
- maintenance and repair	7	7	7
Exterior maintenance	34	37	38
Interior maintenance	21	24	26
Other product finishes	82	84	80
Totals	470	460	418

Source: SRI estimates

based, industrial coatings.   Polymer producers are
advertising acrylic latexes for coil coatings and water-
based urethanes for automobiles and major appliances,
as well as water-based alkyds for metal castings.   The
candidate polymers which might be replaced by a sucrose-
containing polymer are listed (Table IX) by SRI in
more than a dozen categories for all classes of paints.
    Alkyds' "death" has been predicted prematurely for
years by the advocates of competitive polymers.   Their
vital tenacity in the coatings market can be attributed
to, (a) the large numbers of coatings formulators and
polymer synthesizers who understand them well, (b) the
many alternative components for their syntheses, and
(c) the nearly infinite potential for hybridization
with other polymer types such as polystyrene and the
acrylic polymer families.   These points are illustrated and implied in SRI's tabulation (Table X) of alkyd
precursors.

Table IX.   Estimated Consumption of Resins in Paints
and Coatings

	Millions of Pounds		
Resins	1973	1974	1975
Alkyd	760	750	695
Acrylic	363	384	350
Vinyl	413	395	367
Epoxy	94	112	102
Urethane	69	67	60
Amino	74	74	63
Cellulosic	63	56	50
Phenolic	29	27	24
Chlorinated rubber	13	13	14
Styrene-butadiene	30	25	24
Polyester	20	23	18
Natural	22	21	19
Linseed oil	84	58	45
Other	183	194	184
Plasticizers*	49	45	43
Totals	2,266	2,244	2,058

* Plasticizers are sometimes categorized as additives;
however, they have been included with resins in this
report.

Source: SRI estimates.

   Beyond alkyds and closely associated polyesters
(oil-free), other polymer classes such as acrylate esters, amine-aldehydes and urethanes are enjoying much
attention in pioneering new polymers for paints. Some
of this is implied in the other papers which are com-
panions to this one and advocate the use of sucrose in
urethane foams where all eight hydroxyls in sucrose
are chemically functional to achieve high molecular
weight and rigidity. Coatings require more flexibility and sucrose's functionality needs to be modulated
down to a range of two to three for polymer chain formation. In alkyds, the pioneer researchers in use of
sucrose, as reported in papers by Bobalek and Faulkner,
have found necessary the fatty acid esterification of
at least four of the eight hydroxyls of sucrose in order to make the sucrose segment of the polymer suffient-
ly hydrophobic and in order to get enough diene groups
in one alkyd molecule for fast enough film drying by

Table X. Consumption of Specific Alkyd Resin Precursors, 1975

Millions of Pounds

Polybasic acids and anhydrides:			265
Phthalic anhydride	225		
Isophthalic acid	23		
Maleic anhydride	6		
Fumaric acid	4		
Other polybasic acids and anhydrides [a]	7		
Polyhydric alcohols:			160
Glycerin	70		
Pentaerythritol	70		
Other polyhydric alcohols [b]	20		
Fatty oils and fatty acids:			275
Vegetable oils and fatty acids		225	
Soybean oil	140		
Linseed oil	40		
Castor oil	20		
Other vegetable oils [c]	25		
Tall oil fatty acids		40	
Marine oils		5	
Synthetic fatty acids		5	
Modifiers:			65
Resins		15	
Phenolic	13		
Other resin modifiers [d]	2		
Monomers		30	
Styrene and vinyl toluene	25		
Acrylic (methyl methacrylate)	5		
Monobasic organic acids and esters		20	
Rosin and rosin esters	15		
Benzoic acid and other organic acids and esters [e]	5		
Total			765

a. Includes trimellitic anhydride, chlorendic anhydride, short-chain aliphatic dibasic acids, and dimerized fatty acids. Typical short-chain aliphatic dibasic acids include adipic acid, azelaic acid, sebacic acid, and succinic acid.

b. Includes ethylene glycol, diethylene glycol, propylene glycol, dipropylene glycol, butylene glycol, neopentyl glycol, trimethylolethane, trimethylolpropane, and other miscellaneous glycols and polyols.

c. Includes coconut, oiticica, safflower, sunflower, and tung oils.

d. Includes silicone and polyamide resins.

e. Other organic acids and esters include dimethylol propionic acid and the glycidyl ester of a synthetic saturated tertiary carboxylic acid.

Source: SRI estimates.

oxidation or heat.   While these sucrose-containing
alkyds are developed further and exploited, there is
more need now for other routes to the use of sucrose in
paints.

Innovation in nonfood, chemical uses of sucrose
requires understanding of the fundamental chemistry of
sucrose, beginning with more research on the kinetics
of reaction rates of the three primary hydroxyls versus
the five secondary ones.   Using the relative stability
of cellulose as a model, with its one primary and two
secondary hydroxyls per glucose unit, an ingenious chemist can strive toward a goal of synthesizing an oligomer for superior paints by reacting just two of the
primary hydroxyls of sucrose to form a polymer chain of
low to intermediate molecular weight, while reserving
the third primary hydroxyl for in situ crosslinking
during film formation.   These reaction temperatures
need to be well below the caramelization temperature
range of sucrose.

Summary

Prospects for sugar in surface coatings can be
bright for an ingenious chemist who can bring together
the two large stores of carbohydrate and polymer science toward new oligomers for coatings.   His success
will depend critically upon support by a team effort
encompassing process engineering and marketing skills,
among others, and most of all upon the support of a
"champion".

Abstract

Prospects for sucrose in surface coatings will
widen as coatings become even more specialized and as
natural products, including sucrose, regain economic
opportunity to compete against higher priced petrochemicals in coatings.   Finishing in factories rather than
on site has an increasing share of the coatings market
and enables more sophisticated curing mechanisms to
yield more durable coatings.   These advances thus far
have depended mainly on great and complex achievements
in fifty years of petrochemistry toward carefully designed and durable polymer structures.   Maximum use of
chemical reactivity before and during polymer and film
formation leave less vulnerability to degradation of
the film, e.g., 25 to 100 micrometers thick for up to
twenty years.

Sucrose ester developments over 30 years have several established uses in solution type coatings, but

need extension into specialized factory applied coatings of both conventional and newer water dispersible types. Urethanes based on alkoxy modifications of sucrose need extension beyond their use in rigid foams to surface coatings where some types of urethane coatings have superior toughness. Sucrose as a pure, low cost chemical is promising for much exploratory research beyond present esters and ethers to utilize its unique, complex reactivity towards new kinds of polymer formation. This exploration must be guided by changing balances in costs of materials, energy and environmental controls.

Literature Cited

1. Glaser, Milton A.,"Innovation in Organic Coatings", The 1974 Mattiello Memorial Lecture, J. Paint Technology, (1974), December, 39-58.
2. Dean, John C. et al.,"NPCA Data Bank Program-1976", compiled for the National Paint and Coatings Association, by the Stanford Research Institute, Menlo Park, California, USA.

Biographic Notes

John C. Weaver, Ph.D., Retired, consultant to the chemical industry. Educated at Dennison Univ. (Ohio), and the Univ. of Cincinnati. Joined the Sherwin-Williams Co., in 1936, specialities resins and polymers, retired in 1974 as Dir. of Res. Coatings Group. Consultant to educational and industrial organizations. 3305 Enderby Road, Shaker Heights, Ohio 44120, U.S.A.

SAIB in Coatings

CHARLES H. CONEY

Eastman Chemical Products Inc., Kingsport, Tenn. 37662

A systematic synthesis of sucrose esters began at the Research Laboratories of Tennessee Eastman, in 1956. Simultaneously, evaluation of these preparations as potential commercial products with special emphasis as components of surface coatings was conducted by the Technical Service and Development Laboratories of Eastman Chemical Products. From this cooperative effort, success came in the form of a very unusual and unique compound, the mixed acetate-isobutyrate ester of sucrose - SAIB.

The preparation of these esters is fairly straightforward (1). Sucrose and an excess of the anhydride are heated in the presence of the corresponding sodium or potassium salt. The crude ester then is dissolved in hexane and washed with dilute aqueous sodium hydroxide to remove the residual acid, catalyst and color. The hexane is then stripped under reduced pressure.

At the beginning of this study, the octaacetate ester was a commercial product, but the other sugar esters were not commercially available. Thus, the octasubstituted esters of propionic, butyric, isobutyric, valeric, 2-methyl butyric, and 2-ethyl hexanoic acids were prepared (Table I). It was found that, although the octapropionate ester is a glassy solid in a super-cooled state, it soon crystallizes. Sucrose octaisobutyrate is a very viscous liquid after melting but, within a few hours, begins to crystallize. The butyrate, the valerate, the 2-methyl butyrate, and the 2-ethyl hexanoate octa esters, each existed as liquids which did not crystallize. It is interesting to note that the octaisobutyrate ester on crystallization forms a very symmetrical, spherical mass as individual crystals radiate from the nucleus.

Table I.

Sucrose Ester Modified Cellulose Acetate Butyrate Films
(50/50 Ratio)

Type of Octa-Ester	Appearance
Acetate	Very brittle
Propionate	Hazy, brittle
Isobutyrate	Opaque, brittle
Butyrate	Soft, tacky
Valerate	Soft, tacky
2-Ethyl hexanoate	Very soft film

When the sucrose esters were evaluated as modifiers for cellulose acetate butyrate, the crystal-pro-
ducing compounds gave brittle and sometimes hazy or
opaque films as they continued to crystallize, even in
the presence of the cellulose polymer (Table II).   On
the other hand, the liquid esters performed as plasticizers for cellulose acetate butyrate, producing soft
and tacky films at 50% modification.   Because of this
softening action, sucrose esters, forming low-viscosity
liquids, were eliminated from commercial consideration.
It also was estimated that these esters could not compete costwise with the commonly used, coatings plasticizers.

Table II.

Sucrose Esters

Type of Octa-Ester	Physical Nature	Melting Point, °C
Acetate	Crystalline	86
Propionate	Crystalline	45
Isobutyrate	Crystalline	64
n-Butyrate	Liquid	—
Valerate	Liquid	—
2-Methyl butyrate	Liquid	—
2-Ethyl hexanoate	Liquid	—

To overcome the dilemma either of crystalline or
low-viscosity compounds, the mixed esters were produced and investigated.   Several of those were found
to produce viscous liquids which would not crystallize.
From evaluations in combination with various film-forming polymers, sucrose acetate isobutyrate was selected
as producing properties most desirable as a coatings
modifier (Figure 1).   Of course, there are many possible combinations from the various ratios of acetyl to
isobutyryl groups.   From a study of this factor, it

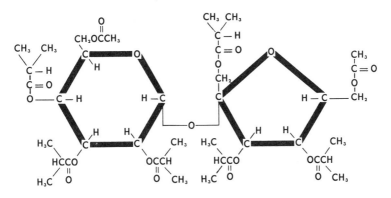

Figure 1

was determined that between 2 and 3 acetyl, and 5 and 6 isobutyryl groups gave most desirable properties.   In addition, the random variation of the arrangement of the two substituent groups increases molecular inhomogeneity and reduces the tendency to crystallize.

This high-viscosity, resinous, mixed acetate-isobutyrate ester was found to modify cellulosic-based coatings in a way that could extend the film-forming polymer, giving higher solids without appreciably lowering coating film hardness (2).   This aspect and several other desirable properties of the ester will be covered in the remaining discussion.

Although this ester performs in several ways like

Figure 2

Figure 3

a polymeric material, it is actually a large, bulky
molecule of about 834 molecular weight (Figure 2).
Of course, it is essential that most coatings modifiers
have a degree of resistance to water. As mentioned
earlier, the esterification of all of the hydroxyl
groups on sucrose changes its hydrophilic-lipophilic
balance (HLB) and surface energy to a high degree.
This radical change probably is due, in part, to the
high density of reaction sites available on sucrose and
to the high degree of conversion to the ester.
     The hydrophobicity of SAIB can be illustrated by
its high contact angle with water. The initial angle
of contact was found to be 110-115°. The product is
also very stable in the presence of water. We found,
in the early work, that, as one might expect,the degree
of hydrolytic stability of the mixed ester increases as
the ratio of isobutyryl to acetyl increases. When
immersed in boiling water for four days, it hydrolyzed
to the extent of only 0.3% by weight.
     The high viscosity of the acetate-isobytyrate ester
(Figure 3) may be reduced dramatically by an increase
in temperature (Figure 4). As an example, its viscosity of greater than 100,000 cps at 25°C can be reduced to 1,000 cps at 68°C, or to 100 cps at 100°C.
At elevated temperatures, the ester loses its inactive
role as modifier-extender and functions as plasticizer
or even solvent for some polymers. This bifunctional
property makes it a valuable component of heat-seal
coatings and hot-melt adhesives.

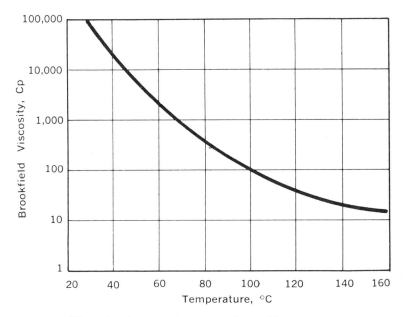

Figure 4. Viscosity of SAIB as influenced by temperature

In today's need to reduce solvent emissions, high-solids coatings have high priority. Two properties of the product suit it particularly well as a modifier for high-solids coatings; these are: low solution viscosities, and minimum effect upon coatings hardness. One can see the viscosity of an acetate-isobutyrate solution is greatly affected by slight changes in concentration at high levels of the product, but is relatively unaffected at low or medium levels (Figure 5). A 100-fold reduction of viscosity is produced by the addition of 10% solvent. Likewise, fairly high levels of the ester may be used with some polymers without appreciably affecting film hardness (Figure 6). As an example, up to 50% SAIB may be used with cellulose acetate without causing a great change in film hardness. Also, 50% SAIB will modify cellulose nitrate to produce a surface hardness greater than that of the unmodified film. Therefore, because of this film hardness effect and the low solution viscosity effect, this ester finds one of its principal applications as an extender to increase coatings solids.

SAIB has been used for a number of years in coatings and saturants for the transparentizing of paper (Figure 7).

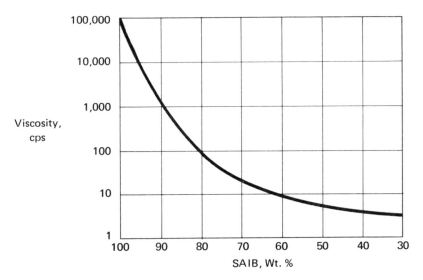

Figure 5.   Solution viscosity of SAIB in typical coatings solvents

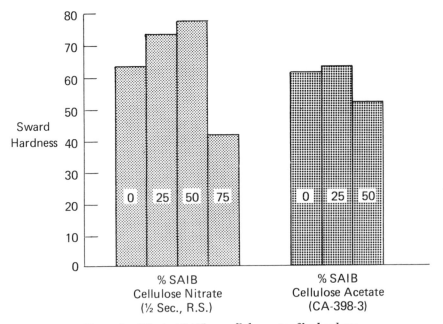

Figure 6.   Effect of SAIB on cellulose ester film hardness

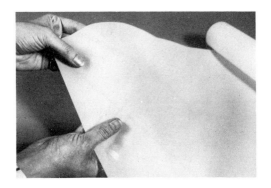

Figure 7

Three properties, which I have not yet discussed,
make this ester especially suited for that purpose.

First, and aside from its good clarity and
low color, the product has a refractive index
close to that of cellulose fiber (3) (Table III).
Thus the individual fibers of a paper tend to
"disappear" as the ester surrounds and reduces
the light reflected from the fiber's surface.

Table III.

Refractive Index $n^{20°}$

SAIB	-	1.454
COTTON	-	1.555
JUTE	-	1.536

Secondly, SAIB has low volatility even at
elevated temperatures (Figure 8) and thus
provides a very permanent degree of transparency
to the paper. Compared with dioctylphthalate
and poly-α-methylstyrene, sucrose acetate-isobutyrate has much less weight loss at temperatures as high as 171°C.

Thirdly, sucrose acetate-isobutyrate is
resistant to discoloration on exposure to
heat and ultraviolet light, adding to the
quality of the transparentized paper.
Despite its liquid form, when used at the
correct level, the ester gives a dry feel
to the paper.

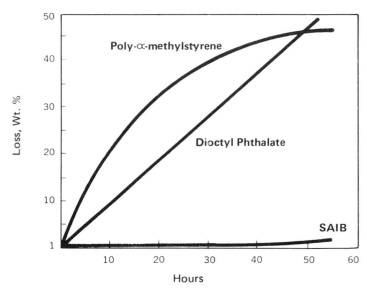

*Figure 8. Volatility of poly-α-methylstyrene and dioctyl phthalate at
350°F*

The high viscosity and wetting characteristics of
the product make it an efficient dispersant for coatings pigments (4). The dispersion, which may contain
various amounts of solvent to control viscosity, forms
a stable suspension which is compatible with many coatings systems and which can be used to tint these
systems. The ester can be used as the medium in
practically all of the pigment dispersion techniques.
These include the three-roll mill, sandmill, ballmill,
attritor, and the process of flushing of wet pigment
cake (5).

In the wet pigment cake flushing method, sucrose
acetate-isobutyrate has been found to be much more efficient than other media such as alkyds and oils ordi-
narily used for this purpose (6). Very likely, the
properties of viscosity, hydrophobicity and the pigment
wetting characteristics of this ester, combine to
separate the pigment from the water more rapidly and
to a higher extent.

Sucrose acetate-isobutyrate may be emulsified
readily with a surfactant mixture having an HLB value
of 14, Table IV. The ester is heated to about 70°C and
an inversion technique is used to produce an oil-in-
water-type emulsion having excellent stability. As
interest increases in water-based coatings as a means
of reducing air pollution, it is significant that SAIB
emulsions may be used in latex-based coatings (5).

Table IV.   SAIB Emulsion Formation

Ingredients	Weight %
SAIB	40
Surfactant	5
Water	55
	100

The product has the desirable property of imparting adhesion to several plastic surfaces.   It can increase adhesion of lacquers (4) to molded or extruded cellulose acetate, cellulose acetate butyrate, nylon and acrylics, and to cellophane and Mylar film. SAIB is compatible with a wide variety of polymers, resins, plasticizers, oils, and waxes.   Thus, the sucrose ester may be used in many types of coatings for various applications both for indoor and exterior exposure, including wood coatings, metal coatings, cloth and paper coatings, plastic lacquers, printing inks, and heat-sealing adhesives.   Many suitable formulations have been developed for these applications (7,2).

As one can see, SAIB is a very versatile compound for use in coatings. Its versatility extends into several other fields.   It is used as an additive in plastics extrusion to obtain improved milling properties and increased surface hardness (8).   A special grade of this ester is used in many countries as a soft drink modifier for flavoring-oil suspension and clouding purposes (9).   It is also reported to be used as an ingredient in explosives, adhesives, polishes, cosmetics, photographic film (10), and perfumes.

A material as unique as this sugar ester is certain to find use in many yet unrealized applications.

Abstract

The physical nature of the acetic, propionic, butyric, isobutyric and valeric esters of sucrose ranges from crystalline solids to low-viscosity liquids.   As coatings modifiers, the crystalline compounds have a tendency to form crystals within the film while the liquid compounds cause the film to soften as a plasticizer would do.   In the search for a coatings modifier which would perform as an extender, mixed esters were investigated.   It was discovered that the completely esterified mixed acetic and isobutyric ester of sucrose produced some unusual properties alone and also when combined with various film formers.   A major aspect of sucrose acetate isobutyrate (SAIB) is its capability of extending polymers to

222

impart various properties without degrading the physical toughness and hardness of the coating, unlike many
other modifiers. The low viscosity of SAIB in most
solvents contributes to the attainment of high-solids
coatings. These coatings are utilized in interior
and exterior applications for wood, paper, plastic, and
metal surfaces. This ester is finding use in a wide
spectrum of applications other than coatings, including
adhesives for laminating plastic film, printing inks,
hot-melt coatings, heat reactivated adhesives, transparentized paper, pigment dispersions and soft drinks.

Literature Cited

1. Touey, G.P., Davis, H.E., U.S. Patent 2,931,802,
    April, 1960.
2. Gearhart, W.M., Ball, F.M., U.S. Patent No. 3,076,
    718. February, 1963.
3. Scott, J.R., Roff, W.J., "Handbook of Common
    Polymers," page 134, CRC Press, 1971.
4. Coney, C.H., American Ink Maker, (1966).
    XLIV, (3).
5. Coney, C.H., Draper, W.E., U.S. Patent 3,318,714,
    May, 1967.
6. Coney, C.H., Draper, W.E., Defensive Publication
    745,090, December, 1968.
7. Coney, C.H., American Ink Maker, (1969), XLVII,
    (1). January.
8. Gearhart, W.M., Wilson, E.W., SPE Journal, (1960),
    16, (10).
9. British Patent No. 1,118,019 (Aktieselskabet Co-
    Ro), June, 1968.
10. Ferrania, Belgian Patent 609,231.

Biographic Notes

Charles H. Coney, B.S., Industrial res. chem.
Educated at Univ. of South Carolina. Joined Eastman
Chemical Products, Inc. in 1948; technical service
and development of chemicals for plastics, protective
and decorative coatings. Coatings Chemical Laboratory, Tech. Service and Devel. Div. Eastman Chemical
Products, Inc., Kingsport, Tennessee 37662 U.S.A.

16

Sucrose Benzoate—The Unique Modifier

E. P. LIRA and R. F. ANDERSON

Velsicol Chemical Corp., 341 East Ohio St., Chicago, Ill. 60611

Sucrose benzoate, prepared via our industrial process, has a typical assay of approximately 7.4 benzoyl
groups per sucrose molecule. This is about the optimum degree of substitution for the conditions imposed
upon the reactants. It should be noted that, although
sucrose benzoate is a discrete molecule, it has a molecular weight approximating that of many oligomeric
systems. Figure 1 depicts sucrose benzoate. Please
note that the outer surface, or shell, is composed
either of aromatic benzene rings or ester groups.
This then is the type of surface which will interact
with the conditions imposed upon it.

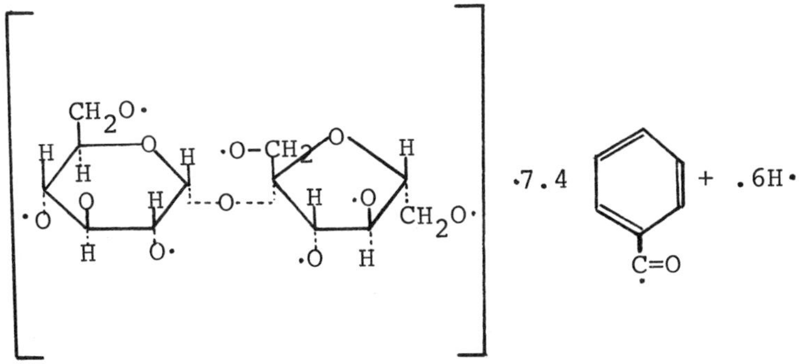

Average Formula $C_{63.8}H_{52.6}O_{28.4}$

Average Molecular Wt.   1111.4

Figure 1. Sucrose benzoate—the unique modifier

There is one significant article in the recent
literature on the preparation of sucrose benzoate (1).
Ness and Fletcher described the preparation of the
octabenzoylated derivative by use of benzoyl chloride
in pyridine, followed by a work-up procedure which
involved a carbon tetrachloride-methanol treatment.
This resulted in a crystalline adduct containing
sucrose benzoate and two moles of carbon tetra-
chloride.   All previous and subsequent reports of
sucrose octabenzoate describe only an amorphous product.   This Ness-Fletcher adduct returned to the
amorphous state after removal of the carbon tetrachlor-
ide under vacuum.   This amorphous state is a second
important characteristic of sucrose benzoate.
    In addition to the above method of preparation,
attempts have been made without success to make benzoic
acid react directly with sucrose.   The process which
this paper outlines follows the classical Schotten-
Baumann technique as illustrated in Figure 2.

Manufacturing Process
(Schotten-Baumann Technique)

Sucrose (Aqueous)
Caustic Solution   Ambient
Benzoyl Chloride   ─────────────▶ Sucrose Benzoate
Toluene   Temperature   in Toluene

                                               │ - Toluene
                                               ▼

                                        Sucrose Benzoate

Process Advantages: Lower Color
                      Higher Yields
                      Less Contamination

Figure 2.   Manufacturing process (Schotten–Baumann technique)

The raw materials are pumped into a reactor and stirred
vigorously, since the system is heterogeneous.   After
a suitable reaction period, the phases are separated
and the toluene removed from the sucrose benzoate.
The solvent-free product then is prepared in a flaked
form.   This process yields a product which has less
contamination and color than other methods and also
gives consistently good yields.
    Tables I and II list some typical physical proper-
ties and/or specifications for the product of the above
process.   As indicated before, it is an amorphous
material which melts over the 95-101°C range.   Ash
and acidity are low and its flash point high.   Its

A.P.H.A. color is very low, in fact, the product is water-white.   Other data of interest, and which strongly influence the use patterns, are the refractive index, UV absorbancy and viscosity.   Especially note that the viscosity in a toluene solution does not increase rapidly until the concentration is quite high.

Table I.
Typical Physical Properties

Form	Non-Crystalline Solid
Softening Point (ASTM E28-67)	95-101°
Sp. Gr. 25°/25°C	1.25
Ash (ASTM E347-71)	0.01% maximum
Flash Point, TOC	500°
Acid Value, Mg. KOH/G.	Less than 0.1
Hydroxyl Value, Meg./Gm.	0.9
Saponification Equivalent	150
APHA Color (ASTM D1209-69)	30
N$_{25}^{D}$	1.577
λMax	230mμ
100% Transmittance	>300mμ

Viscosity (25°C) in Toluene		
25% Solids	1.30	Centipoises
40% Solids	3.7	Centipoises
60% Solids	35.1	Centipoises
70% Solids	530.0	Centipoises

The stability profile data in Table II, generally are quite favorable.   Sucrose benzoate is resistant to prolonged treatment with boiling water and has surprisingly good stability against both acid and alkaline conditions.   The rate of hydrolysis appears initially to be faster in base, but then may turn out to be slightly slower than in an acid treatment.

Table II

Stability Profile

Hydrolysis (100°C)

	24 hours	96 hours
Water	0	0
5% Aq. HCl	0	0.6%
5% Aq. Na$_2$CO$_3$	1.9%	2.3%

Thermal stability was followed by color change and increase in acidity as shown in Table III. The rate of decomposition is very slow at 150°C. Even at 175°C, in an aluminum dish, the increase in acidity was only from .001 meg H+/g to .007 meg H+/g. It has been found that epoxy-containing additives can increase the heat stability of the product.

Table III

Stability Profile

Thermal (150°C)

Hours	Color (APHA)	Meq.H⁺/g
0	30	.0025
6	70	.004
24	250	.009

UV Light (140°F)

(Atlas Fade-O-Meter)   Film (4 mil)
                         - No color change after
                           3000 hr
                         - Infrared unchanged after
                           3000 hr

UV stability studies indicate virtually no change in the chemical constitution of sucrose benzoate when it is subjected to prolonged exposure to a filtered, carbon arc light source at 140°F. The principal maxima of this light source is between 340 - 440 mμ. As noted previously, the λmax. of sucrose benzoate is around 230 mμ which is well below the wavelength which will penetrate air. It is completely transparant above 300 mμ.

As might be expected of a structure with aromatic and ester character, the solubility and compatibility properties have appreciable latitude, as shown in Table IV. Sucrose benzoate is virtually insoluble in hydrocarbons and water. It has low solubility in lower molecular weight alcohols and glycols. Only the olefinic polymers are not compatible, while substances with a large hydrocarbon portion might be only compatible at low levels (e.g. 2%) (See Table V). Examples of this behavior are cottonseed oil, corn oil, stearic acid and tall oil fatty acids. This broad compatibility makes sucrose benzoate a potentially useful modi-

fier in many systems and can be used to cause "mutual"
compatibility between "marginally" compatible systems.

Table IV.

Solubility

Water (20°C)	.001%
(67°C)	<.01 %
Heptane (67°C)	.02 %
Ethyl Alcohol	2.3 %
Most Acids, Amides, Aromatics, Ethers, Ether Alcohols, Esters, Halocarbons, Ketones and Nitriles	Completely Miscible
Alcohols, Glycols	Sparingly Soluble

Table V.
Compatibility

Polyethylene, Polypropylene and Polyisobutylene	I
PVA, PVC/PVA	C
Polystyrene	C
Polyester	C
Ethyl Cellulose, Nitrocellulose and Cellulose Acetate-Butyrate	C
Acrylic	C
Alkyd Resins	C
Nylon	C
Urea-Formaldehyde, Melamine-Formaldehyde	C

A summation of some of the properties which may
allow sucrose benzoate to be called a "unique modifier"
are listed in Table VI.   Although most have been
mentioned previously, the gloss as well as clarity and
lack of color obtained from the use of this material is
outstanding.   Its amorphous nature eliminates any concern about crystallization in the modified system.   It
has excellent hardness, but in many applications it
needs to be plasticized because of its brittleness.
      Table VII presents the major uses for this product,
that is, before an unrelated explosion and fire in the
plant caused Velsicol to stop production.   At that
time a production of nearly a million lb/yr was prepared and sold.   Within the past few months Velsicol

has just reinstituted production on this product and so
sucrose benzoate is available at this time.

Table VI.

Unique Characteristics

I.	Excellent UV Stability
II.	Low solution viscosity over large range of concentrations.
III.	Imparts excellent gloss
IV.	Unusual clarity and waterwhite color
V.	Non-crystallinity
VI.	Hardness
VII.	Improved pigment wetting and dispersion rates
VIII.	Excellent adhesion in plasticized system
IX.	Broad range of compatibility

Table VII.

Major Uses

Lacquers
  I.   Nitrocellulose
       — Blend - 60% sucrose benzoate/40% DOP
       1.   Lower solution viscosity
       2.   Better clarity
       3.   Better alcohol spot resistance
       4.   Better UV yellowing resistance
       5.   Hardness with flexibility
       6.   Limits DOP exudation

  II. Cellulose Acetate Butyrate
       — Blend - 75% sucrose benzoate/25% DOP
       Above described characteristics

  III. Acrylics
       — Blends with plasticizers
       Above plus improved pigment
       wetting and dispersion rates

  IV. Polyvinyl Chloride-Acetate
       — Blends with plasticizers
       Above plus improved resistance to
       acids and alkaline material

The major marketing area was in lacquers and most
of the previously described properties make it well

suited for this market.   Again, note that in these
applications a blend of sucrose benzoate with a plasticizer is a must.   Our results indicate that the exuda-
tion from this system was significantly retarded.   In
the case of nitrocellulose, the blend described has the
advantages listed in Table VII when compared with conventional coconut alkyds.   Similar properties also are
available in the other three systems.

   Finally, in Table VIII, we have a summary both of
minor uses and reported potential uses.   Most of these
are dependent on sucrose benzoate as a viscosity modifier.   However, as noted in items C and D, when
applied in an appropriate manner to cellulose, it
causes the cellulose to become "transparant".   This
apparently is due to the similarity in refractive in-
dices.   The second use, the dry cleaning size application, is the result of the lubricity of the material
when deposited on the fiber.   Its use in this area
also is influenced by lack of odor, color and perma-
nence.   As an adhesive component, it was used to bind
the paper to the filter of Parliament cigarettes.

Table VIII.

Minor and/or Potential uses

A. Adhesives - Viscosity Modifier
   FDA listing (CFR pp. 121, 2520)
B. Plastisol - Viscosity Stabilizer
   PVC, DOP, Sucrose Benzoate
C. Paper Transparentizing
D. Dry Cleaning Size
E. Aminoplast Molding Composition - Flow
   Promoter
F. Electroscopic Toner Powders -
   Viscosity Modifier Melting Characteristics

Abstract

   Industrially prepared, sucrose benzoate is a partially esterified sucrose with a degree of substitution
of approximately 7.4.   Preparation is via the classical Schotten-Baumann technique utilizing two immiscible
solvents which results in better yields with increased
purity over other techniques.   Sucrose benzoate has
excellent UV stability, unusual clarity and lack of
color, a low solution viscosity, yields films of excellent depth or fullness of gloss and has a broad range
of compatibility with numerous resin systems.   It is

thermally stable and resistant to hydrolysis under
weakly acid or alkaline conditions. It has FDA approval as a component in food-packaging adhesives. Based
on available information, sucrose benzoate would be
regarded neither as a "highly toxic substance" nor as a
"toxic substance". Systems which have benefited from
the use of sucrose benzoate as a modifier are nitrocellulose lacquers, cellulose acetate butyrate lacquers, polyvinyl chloride-acetate coatings and acrylic
lacquers.

Literature Cited

1.   Ness, R.K., and Fletcher, H.G., Journal of the
     American Chemical Society, (1952), 74, 5344-46.

Biographic Notes

     Ralph F. Anderson, Ph. D., Vice President - Re-
search. Educated at the Univ. of Wisconsin. Biochemist U.S. Dept. of Agriculture; Dir. Res. and Vice
Pres. R & D, Minerals and Chemicals Corp., in 1973 became Vice Pres. of Res. at Velsicol Chemical Corp.
341 E. Ohio St., Chicago, Ill. 60611 U.S.A.

Biographic Notes

     E. Patrick Lira, Ph.D., Dir. Res.   Educated at
Elmhurst Coll. (Illinois), and Rutgers, The State Univ.
(New Jersey).   Joined IMC Corp. in 1963 and Velsicol
in 1973.   Organic chemist, specializing in pesticides
and industrial chemicals.   Research Dept. Velsicol
Chemical Corp., 341 E. Ohio St., Chicago, Ill. 60611
U.S.A.

Discussion

Question: Professor Bobalek, would you comment on the dryness of sucrose, the anhydrous conditions, and the practicality of the reaction?

Professor Bobalek: If everything else is dry, the solvent and the methyl fatty esters, the reaction can tolerate rather modest anhydrous conditions in the sucrose. Drying overnight at a few degrees above the boiling point of water is enough to prevent trouble. We even have used sugar right from the bag, if everything else was dry, with no noticeable differences in results. However, the moisture problem gets more and more aggravated, of course, as both the scale and the degree of reaction advance. One has reason to be more concerned about preventing invasion of water which would complicate and slow the reaction at D.E.s above 3.5 under the realities of the reaction on a plant scale.

Question: Professor Bobalek and Mr. Faulkner, would you comment on the preliminary economics of the two approaches to surface coatings you just mentioned?

Professor Bobalek: If one wants to make a detailed cost analysis, he runs into a rather sticky problem. First,one must establish the bench marks. In this case let us agree that sucrose esters are going to be a commodity chemical, to be produced in an amount of 10 million lb/yr minimum. The linseed oil market is in the order of a half a billion lb, worldwide. The alkyd grade of tall oil acids is around 81 million lb. Thus, we are talking of a major invasion of the paint business with these sucrose esters as commodity chemicals. To become a major commodity, in five years, sales would have to achieve a level of 200 million

lb minimum in order to sustain the capital investment
in the costs of this kind of plant with a 6 to 10 year
payoff. The price will demand that the capital burden
represent no more than 3% of the poundage costs of the
product. If they are to sell, the cost of the products probably will have to average 60¢ maximum, in order to cover a spectrum of different versions. Cost
will range from 25¢ to $1.50, depending on whether
they are aimed at an exalted domain such as printing
inks or sausage casings, or used as crude varnish to be
thrown on the bottom of car fenders.

Mr. Faulkner: I endorse what Professor Bobalek
has said. I have little help to offer on economics
because we really have not gone into the economic aspect. All our guestimates on the price of sucrose
resins have been based on raw materials costs. Our
main objective was to demonstrate that we could form a
satisfactory paint resin from sucrose drying oil esters
and this has been achieved using "penta" esters. In
the case of our first resins of the epoxy ester type,
it was clear that the price, based on raw materials'
costs, would be too high in terms of the performance
obtained. However, sucrose polyurethane resins certainly are in the right ball-park and it appears more
likely that, with this type, performance should match
their price.

Question: Dr. Lira commented on the UV stability
of the sucrose benzoate. I wonder if that imparts UV
protectivity to the PVC resins where there exist such
problems?

Dr. Lira: If you are asking whether sucrose benzoate as a modifier in PVC would impart UV stability,
the answer is, no. Sucrose benzoate contains no chro-
maphor through which to distribute the energy.

Question: Dr. Lira, have you worked with glucose
esters?

Dr. Lira: Glucose esters are among those sugar
esters we have looked at without, what I would call, a
resounding success. There appear to be some problems
in the manner in which we handle it.

Question: There would seem to be approximately
one hydroxyl group unsubstituted in your sucrose benzoate. Do you know where this hydroxyl group resides?
In other words, which is the least readily esterified

hydroxyl group, and have you considered further modifying your products, for example, by introducing an isobutyrate grouping on the unsubstituted hydroxide?

Dr. Lira: In response to the first part of your question, we do not know which one is the unsubstituted hydroxyl.   To continue with your question, we have not considered exhaustive substitution on that last 0.6-hydroxyl function,but we have considered modifications, and are in the process of working on some of them, involving long chain substitutions as partial replacements for some of the benzoyl functionality.

Urethanes and Fermentation Sucrochemistry

Introduction

L. HOUGH

Department of Chemistry, Queen Elizabeth College, Atkins Bldg., Campden Hill, Kensington, London W8, England

If we consider the various applications of sucrose, there can be little doubt of the enormous potential in the plastics industry. On theoretical grounds
the multi-functionality of sucrose would appear to be
well suited to production of urethane foams. Thus,
whilst appreciating that sucrose has been incorporated
into the important phenolics and melamines, the International Sugar Research Foundation has elected to high-
light industrial applications in the urethane field for
two excellent reasons - not only has there been phenom-
enal growth in this area, but sucrose has gained general acceptance for incorporation into rigid urethane
foams.

An equally important area for the utilisation of
sucrose lies in the wide variety of fascinating products that can be produced by fermentation, ranging
from the well-known wines and spirits to the antibiotics. Currently three diverse applications of fermentation technology appear to be ripe for development,
namely the economic production of organic solvents (because of the rising cost of petrochemical sources), the
formation of the gums - complex polysaccharides - that
have properties suiting them for scientific industrial
applications and finally, the conversion of sucrose or
molasses into single cell protein to meet the ever-in-
creasing demand for this foodstuff.

It is important to recognise that our speakers'
contributions have considerable relevance, in the sense
that applied carbohydrate science will assume increasing importance to Society as fossilized materials are
gradually consumed by that Society to be replaced,
gradually but of necessity, by substrates that are re-
plenishable.

17

An Overview of Sugars in Urethanes

K. C. FRISCH and J. E. KRESTA

Polymer Institute, University of Detroit, 4001 W. McNichols Rd., Detroit, Mich. 48221

Polyurethanes, or urethanes as they are generally
referred to in industry, are among the fastest growing
polymers in the world. Historically, urethanes were
first developed in the laboratories of the I.G. Farben-
industrie in Germany in the nineteen thirties under the
leadership of Professor Otto Bayer, and it was not until the late fifties that urethanes gained significant
industrial importance in the United States. It is the
more remarkable that urethanes grew in this country
from 8 million pounds in 1956 to about 1.4 - 1.5 billion pounds in 1975 (1-3). In order to understand
the continued rapid growth and commercial acceptance
of urethanes in the various markets, one has to realize
that urethanes are probably the most versatile class of
polymers. The principal method of manufacture has
been the reaction of hydroxyl-terminated polyethers or
polyesters (commonly referred to as polyols) with di-
or polyisocyanates which can be represented schematically as follows:

$$HO-R-OH \ + \ OCN-R^1-NCO \ \longrightarrow \ \left[-O-R-O-\overset{\displaystyle O}{\overset{\displaystyle \|}{C}}-NH-R^1-NH-\overset{\displaystyle O}{\overset{\displaystyle \|}{C}}- \right]_n$$

polyether   diisocyanate   polyurethane
  or
polyester

If the functionality of the hydroxyl or isocyanate
component is increased to three or more, branched or
crosslinked polymers are formed. The properties of
urethane polymers are dependent primarily upon molecular weight, degree of crosslinking, effective inter-
molecular forces, stiffness of chain segments, and

crystallinity. Due to the many structural variations
that are possible, urethanes can be formulated and
processed into many diversified forms. They include
flexible, semirigid and rigid foams; soft and hard
elastomers, coatings, and adhesives; thermoplastic and
thermosetting plastics, fibers (Spandex), films; poro-
merics; etc. A breakdown of the principal areas of
urethane polymers in million pounds is shown in Table I
(1). The principal suppliers of urethane raw materials (isocyanates and polyols) and an estimate of
their respective production capacity for 1976 are
listed in Table II (3).
        Sugars and sugar derivatives play very important
roles in the manufacture of urethane foams, especially
of rigid urethane foams, While polyols used for the
manufacture of flexible foams, coatings, adhesives, and
elastomers generally have a functionality (f) of 2 to
3 (2-3 OH groups), polyols used for the manufacture of
rigid foams have usually a functionality of 4 or greater. The most commonly used polyols for rigid foams
are based on sucrose (f=8), sorbitol (f=6), pentaerythritol (f=4) and on aliphatic or aromatic polyamines
(4-5), such as ethylenediamine, diethylenetriamine,
tolylenediamine and condensation products of aniline
and formaldehyde. Due to their superior hydrolysis
resistance and lower costs, polyether polyols have been
employed in preference to polyester polyols although
many flame retardant polyols containing phosphorus and
halogens possess ester linkages.
        The polyether polyols are produced by the addition
of alkylene oxides, primarily propylene oxide, to the
above polyols using mostly base catalysis. The reaction mechanism of the base catalyzed addition of propylene oxide to a polyol (schematically represented as
$(R-CH_2OH)$ can be represented as shown in Figure 1(4-6).
        The basic catalyst forms anions by the action of
the catalyst upon the polyol initiator leading to the
opening of oxirane ring and the formation of a new anion. Propagation occurs by successive attacks of
these anions upon propylene oxide monomer. Chain
termination results by combination of the polymer anion
with a proton. This representation may be an over-
simplification since Gee, et al (7) have shown that ion
pairs may be involved. While the above mechanism
shows only the formation of terminal secondary alcohol
groups, a certain amount (5-10%) of primary alcohol
groups is also formed in the oxirane ring opening.
        The ring opening polymerization of 1,2-epoxides
has been reviewed by Ishii and Sakai (8) and Price (9)
and more recently by Lundsted and Schmolka (10).

Table I.   Overall Urethane Demand - 1975, and Outlook - 1976

                                                    Demand *

                                               Millions of Pounds

                                                  1975   1976
- -
Foams

     Rigid   293   353

     Flexible   915   1043

     SUB-TOTAL   1208   1396

Elastomers   84   94

Surface Coatings   77   82

Adhesives and Sealants   37   42

     TOTAL   1406   1614
* On a urethane resin basis

Table II.   URETHANE RAW MATERIAL SUPPLIERS AND CAPACITIES
                         1976 Estimate
                    (Millions of Pounds)

     TDI

                    Allied   80
                    BASF Wyandotte   90
                    Dow   90
                    Dupont   105
                    Mobay   225
                    Olin   130
                    Rubicon   40
                    Union Carbide   55

                         Total   815

     MDI
                    Jefferson   35
                    Mobay   160
                    Rubicon   60
                    Upjohn   240

                         Total   495

     Polyols
                    BASF Wyandotte   330
                    E. R. Carpenter   120
                    Dow   400
                    Jefferson   100
                    Mobay   180
                    Olin   210
                    Union Carbide   400
                    Other   250

                         Total   1,990

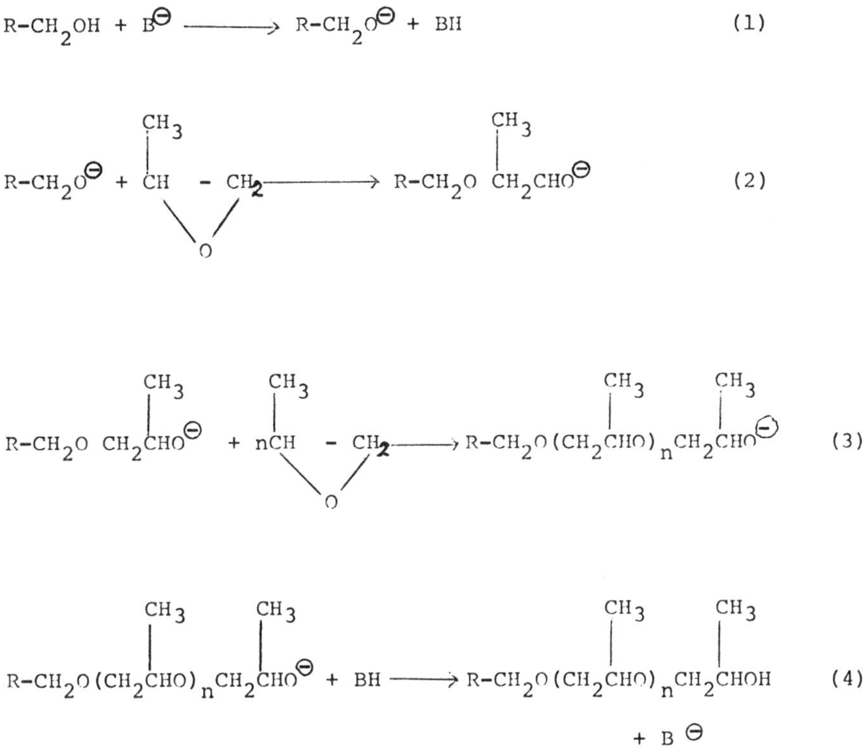

$$R-CH_2OH + B^{\ominus} \longrightarrow R-CH_2O^{\ominus} + BH \qquad (1)$$

$$R-CH_2O^{\ominus} + \underset{\underset{O}{\diagdown\diagup}}{\overset{\overset{CH_3}{|}}{CH} - CH_2} \longrightarrow R-CH_2O\ \overset{\overset{CH_3}{|}}{CH_2CHO^{\ominus}} \qquad (2)$$

$$R-CH_2O\ \overset{\overset{CH_3}{|}}{CH_2}\overset{\overset{CH_3}{|}}{CHO^{\ominus}} + n\underset{\underset{O}{\diagdown\diagup}}{\overset{\overset{CH_3}{|}}{CH} - CH_2} \longrightarrow R-CH_2O(CH_2\overset{\overset{CH_3}{|}}{CHO})_n CH_2\overset{\overset{CH_3}{|}}{CHO^{\ominus}} \qquad (3)$$

$$R-CH_2O(CH_2\overset{\overset{CH_3}{|}}{CHO})_n CH_2\overset{\overset{CH_3}{|}}{CHO^{\ominus}} + BH \longrightarrow R-CH_2O(CH_2\overset{\overset{CH_3}{|}}{CHO})_n CH_2\overset{\overset{CH_3}{|}}{CHOH} \qquad (4)$$

$$+ B^{\ominus}$$

Figure 1.   Mechanism of base-catalyzed addition of propylene oxide to polyols

When terminal primary hydroxyl groups are desired ethylene oxide may be used either as the sole alkylene oxide with the polyol initiator, or it may be employed to "cap" or "tip" the oxypropylated polyol with ethylene oxide as shown in the following scheme:

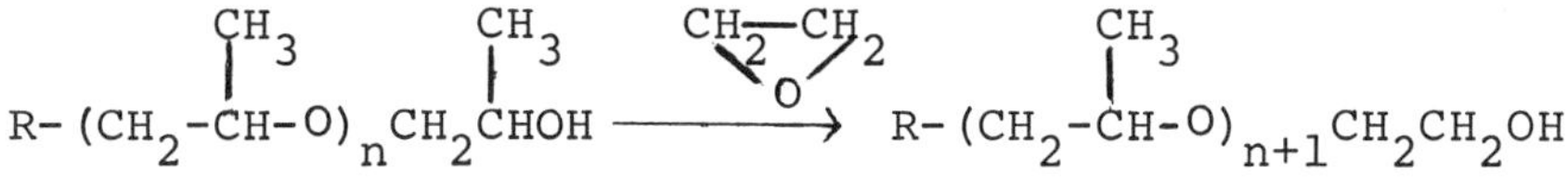

$$R-(CH_2-\overset{\overset{CH_3}{|}}{CH}-O)_n CH_2\overset{\overset{CH_3}{|}}{CHOH} \xrightarrow{\underset{O}{\overset{CH_2-CH_2}{\diagdown\diagup}}} R-(CH_2-\overset{\overset{CH_3}{|}}{CH}-O)_{n+1} CH_2CH_2OH$$

However, since oxyethylene groups are hydrophilic in nature and most applications require good water resistance, polyether polyols for rigid urethane foams consist usually of oxypropylene adducts of polyol or polyamine initiators.   The oxypropylene adducts exhibit also better fluorocarbon (blowing agent) solubility and improved isocyanate compatibility as compared to the corresponding polyols containing oxyethylene adducts.

Base catalysis usually is preferred in the epoxide condensation of "neutral"polyols with potassium hydroxide being the favored catalyst although other alkali hydroxides, alkali alkoxides, and various tertiary amines may be used.

Tertiary amine catalysts are suggested for the addition of alkylene oxides under anhydrous conditions to alkali sensitive, solid polyol initiators such as sucrose (11, 12). It has been claimed that the use of a trialkylamine catalyst containing two or three carbon atoms in the alkyl group limits the addition to only one hydroxypropyl group on each of the hydroxyl groups of the initiator (12).

When halogen atoms are present in the epoxide such as in epichlorohydrin, 3,3,3-trichloropropylene oxide (TCPO) or 4,4,4-trichloro-1,2-butylene oxide (TCBO), or in the initiator, acid catalysts, e.g. boron trifluoride etherate, may be used (13-18). Vogt and Davis (16) found that, if the concentration of catalyst/initiator (polyol) complex is decreased with respect to TCPO in order to obtain higher molecular weight products, side reactions such as cyclization reactions become increasingly important. Boron trifluoride also promotes dimerization of alkylene oxides to dioxane or alkyl derivatives of dioxane as described by Fife and Roberts (19). The use of acid catalysts, e.g. Lewis acids, promotes formation of a greater amount of terminal primary alcohol groups when compared to base catalysis of epoxides.

When uncatalyzed, primary alcohol groups react with isocyanates two or three times as fast as secondary alcohol groups, whereas the presence of catalysts, particularly metal catalysts, cause an even greater spread in reactivity between primary and secondary alcohol groups (20-22). Recently Knodel (23) reported the use of mixtures of ethylene oxide and propylene oxide in the preparation of polyether polyols from solid polyol initiators such as sucrose in the presence of trimethyl- or triethylamine as catalysts. This process was said to reduce the preparation time for the polyether polyols by as much as 67 percent, and the viscosity of the resulting polyether polyol was lower than in conventional processes using propylene oxide as the sole alkylene oxide.

The isocyanates used in rigid urethane foams consist primarily of polymeric isocyanates, i.e. isocyanates having an NCO functionality of greater than 2 (most of them = 2.6-2.8). These are produced by phosgenation of aniline - formaldehyde condensation products and are also referred to as "crude" MDI. The

commercial products from the various isocyanate suppliers differ somewhat in their reactivity which is due
mainly to the ratio of the o- and ρ-NCO substitution
in the molecule, the molecular weight distribution of
the polymeric fractions of the "crude" MDI and also
partially due to the different acidity in the isocyanates (24).   In addition to "crude" MDI, "crude" (un-
distilled grades) TDI (tolylene diisocyanate), also are
employed in the manufacture of rigid foams, especially
for use in appliances such as refrigerators and freezers.

The most important sugars or sugar derivatives
used in the preparation of polyols are sucrose, sorbitol, α-methyl glucoside and dextrose although other
di- and monosaccharides and derivatives thereof have
been employed in smaller quantities.   Due to their
relatively high functionality (4-8), polyols derived
from these initiators find applications primarily in
the manufacture of rigid foams.   Polyols based on
α-methyl glucoside were phased out of the market almost three years ago when CPC International, the major
supplier of these polyols, discontinued their manufacture.   It is estimated that about one third of the
rigid urethane foam polyols are based on sucrose and
sorbitol (25).

Many procedures have been developed for the manufacture of polyether polyols based on the above mono-
and disaccharides as well as some of their derivatives
such as the corresponding polyhydric alcohols (26-44).
Since the oxypropylation of solid initiators, such as
sucrose and other solid polyhydric alcohols particularly for low adducts of propylene oxide (high hydroxyl
number), leads to the formation of polyether polyols
having very high viscosities (as high as >1,000,000 cps
at 25°C), a number of coinitiators are being used to
reduce the viscosity and to impart other desirable properties, e.g. reduced friability, lower combustion,
etc.   These include the use of water (26-29), polyols
such as glycerol (30-32), sorbitol (33), alkanolamines
such as triethanolamine (34,35) and diamines such as
ethylenediamine (36).

The presence of water in any significant amount
lowers the functionality, i.e., the number of hydroxyl
groups in the resulting product, because the polyether
derived from the reaction of water with propylene oxide
has a functionality of 2, thus reducing the average
functionality of the oxypropylated polyether polyol.
High functionality is desirable for good dimensional
stability since it has been shown that dimensional
stability increases with increased functionality of the

polyether.   Therefore, polyols with a functionality of
3 or greater and amine-based polyols are used widely in
the preparation of low viscosity polyols imparting good
dimensional stability to the resulting foams.   Phenol
formaldehyde condensation products (hydroxyl-containing
phenolic resin intermediates) also may serve as coinitiators with sucrose in the manufacture of polyether
polyols for rigid urethane foams (39).   As has previously been pointed out, while propylene adducts of
polyol initiators generally are preferred, combinations
of oxyethylene and oxypropylene adducts have been employed, especially with sucrose as initiator (23, 37,
38).
     In addition to polyols derived from mono- or di-
saccharides or their derivatives, polyether polyols
also have been prepared from starch-derived glucosides.
They generally are prepared by the transglycosylation
of starch with polyols such as glycerol or ethylene
glycol in the presence of an acid (45-49) to yield a
mixture of glucosides as shown in Figure 2.   The
resulting glucoside mixture then reacts further
with propylene oxide to yield the corresponding polyethers.   Different types of polyols may be used in the
transglycosylation reaction as well as glycol ethers of
the general formula ROCH$_2$CH$_2$OH (47).
     In addition to coinitiation of polyols, blends of
polyols are being employed to improve certain foam

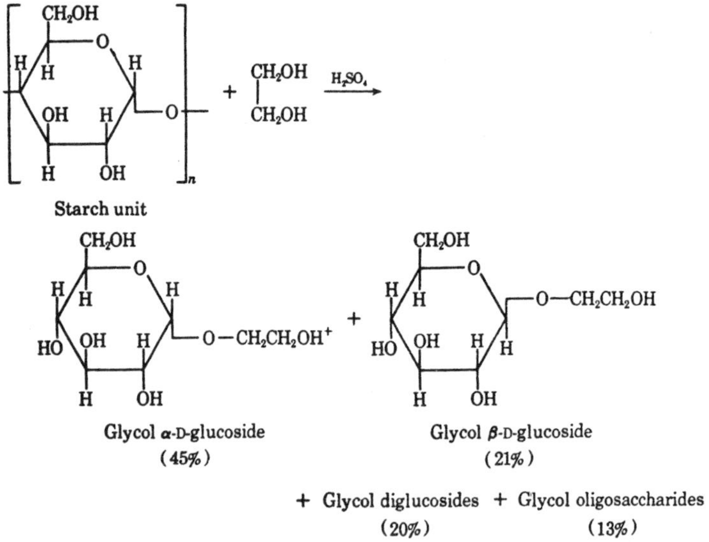

Starch unit

Glycol α-D-glucoside
(45%)

Glycol β-D-glucoside
(21%)

+ Glycol diglucosides + Glycol oligosaccharides
(20%)   (13%)

Figure 2

properties such as friability and flame resistance.
For instance, sucrose and sorbitol-based polyether
polyols frequently are blended with tertiary amine-
containing polyols (e.g. oxypropylene adducts of ethylenediamine or diethylenetriamine) to reduce the friability of the resulting foams.   At the same time,
the presence of the tertiary nitrogen atom in the
amine-containing polyols has a catalytic effect in
promoting faster reaction rates, hence reduces the
amount of extraneous catalyst(s) required.   In addition, lower amounts of flame retardant polyols or
additives are required to attain a certain degree of
low combustion in the foam due to the well known syner-
gism between nitrogen and phosphorus or (and) halogens.
    Blending of sugar-based polyols with flame retar-
dant polyols or additives also is quite common.   Certain flame retardant additives such as tris-(chloro-
ethyl) phosphate also serve as a viscosity reducing
agent although the addition of greater amounts of this
phosphate may affect adversely certain foam properties,
e.g. humid aging characteristics.
    Many flame-retardant polyols based on sugars and
sugar derivatives have been prepared, most of them
containing phosphorus, halogens, or a combination of
these elements (17, 50-59).   An example of a flame-
retardant sugar-based polyol containing only halogens
are adducts of trichlorobutylene oxide (TCBO) to oxy-
ethylated sucrose (17).   The preparation of this TCBO-
containing polyol is shown schematically below:

$$R\text{-}OH + nCl_3C\text{-}CH_2\text{-}CH\underset{O}{-}CH_2 \xrightarrow{BF_3(C_2H_5)_2O} R\text{-}O\left[CH\text{-}CH_2\text{-}O\right]_n H$$

Initiator   TCBO   Catalyst   TCBO-Polyol

Other halogen-containing sugar based polyols include
chlorinated or brominated allyl glucoside polyethers
which have been reported to yield foams of low combust-
ibility and good resistance to dry and humid aging (50-
51).   Hydroxyl-containing tetrabromophthalic esters
(by reaction of tetrabromophthalic anhydride with an
excess of a glycol or another polyhydric alcohol, e.g.
sorbitol) which are then propoxylated, have been used
in the preparation of flame retardant urethane foams
(60).   However, more effective flame-retardant polyols
were obtained by the combination of tetrabromophthalic
anhydride, oxypropylated sucrose, phosphoric acid and
propylene oxide (61).

A number of sugar-based polyols (e.g. derived from sucrose, α-methyl glucoside, dextrose, sorbitol, etc.) containing phosphorus in the form of phosphate, phosphite or phosphonate linkages have been reported for use in flame-retardant, rigid urethane foams (52-59). Recent reviews on flame-retardant urethanes include those of Papa (62) and Frisch and Reegen (63). The effects of chemical structure of polyether polyols for rigid foams and of di- and polyisocyanates on the physical properties of the resulting urethane polymers have been described by a number of authors (64-67). Darr, et al (64) employed Vicat softening points data as a measure of modulus for solid cast urethanes prepared from rigid foam components, but omitting the blowing agent. The polyethers including sucrose, α-methyl glucoside and sorbitol employed in the preparation of the solid polyurethanes are listed in Table III (64). The isocyanates used were HDI (hexamethylene diisocyanate), TDI (tolylene diisocyanate) and MDI (diphenylmethane diisocyanate) with functionalities of 2,2,5 and 3. Table IV lists the Vicat softening points of the solid urethanes derived from the polyols shown in Table III and the isocyanates listed in Table IV. As can be noted from Table IV, the softening points increased as the functionality increased and the equivalent weight decreased. The substitution of a heterocyclic nucleus such as in α-methyl glucoside and sucrose for an aliphatic backbone (e.g. pentaerythritol) in the polyether increased the softening point of the urethane polymer by about 50°C.

The functionality of the polyols also has a profound effect on the properties of rigid foams. Higher functionality favors greater heat resistance and di-

Table III.
TYPICAL CHEMICAL PROPERTIES OF POLYETHER POLYOLS

BASE POLYOL	FUNCTION-ALITY	EQUIV-ALENT WEIGHT	HYDROXYL NUMBER	VISCOSITY, CPS, AT 25°C.	CYCLIC, WEIGHT, %	TYPE
Trimethylolpropane	3	150	400	615	0	aliphatic
Pentaerythritol	4	125	450	1,550	0	aliphatic
Pentaerythritol	4	150	375	1,150	0	aliphatic
Sorbitol	6	125	490	10,000	0	aliphatic
Sorbitol	6	150	380	3,000	0	aliphatic
α-Methyl glucoside	4	125	460	>100,000	16.4	heterocyclic
α-Methyl glucoside	4	150	370	22,500	13.2	heterocyclic
Sucrose	8	125	450	200,000	14.8	heterocyclic
Sucrose	8	150	375	30,000	12.3	heterocyclic
Aromatic Triol*	3	150	380	30,000-40,000	50.0	aromatic

* Not otherwise identified by manufacturer.

Table IV. VICAT SOFTENING POINTS OF HIGHLY CROSSLINKED
URETHANE SOLID POLYMERS (SOFTENING POINTS, °C.)

	RESINS								ARO-MATIC
	ALIPHATIC					HETEROCYCLIC			
Isocyanate	3(150)ᵃ	4(125)	4(150)	6(125)	6(150)	4(125)	4(150)	8(125)	8(150)	3(150)
HDI	22	35	23	60	45	80	63	90	63	52
TDI	79	94	75	136	96	160	126	168	130	112
MDI-2	95	103	99	155	116	178	140	>200	154	127
MDI-2.5	105	120	100	160	124	150	150	—	165	133
MDI-3	110	130	108	170	122	154	160	170	150	150

(a) Numbers in parentheses indicate equivalent weights.

Table V.

1975 U. S. RIGID FOAM MARKET

(Millions of Pounds)

	Pounds	%
Construction	133	39.1
Tanks and Pipe	44	12.9
Appliances	70	20.6
Transportation	36	10.6
Furniture	20	5.9
Marine Flotation	10	2.9
Packaging	12	3.5
Miscellaneous	15	4.4
	340	100.0

mensional stability (assuming that the equivalent
weight is the same) (65). The compressive strength of
the foams usually tends to increase with increased
functionality while the tensile strength and elongation
tend to decrease.

A breakdown of the rigid urethane foam markets for
1975 is given in Table V (3). It can be seen readily
that the construction market represents by far the
largest segment of the rigid foam market. A more recent update of the size of the rigid foam markets in
construction ranged from 145 million pounds (1) to 176
million pounds (68).

Mobay Chemical Corporation marketing representa-
tives estimated that 975 million board feet (209 million pounds) of urethane foam insulation will be used
by the construction industry in 1976, which represents
a 19 percent increase over 1975's 818 million board
feet (176 million pounds) (68). A summary of Mobay

Table VI.
Summary of Mobay Estimates of Urethane Foam Consumption
in Construction Market (all figures are in millions)

	1975		1976		
	Board Feet	Pounds	Board Feet	Pounds	% Change
Pour	144	30	192	40	+33
Spray	159	37	184	43	+16
Board	339	65	407	78	+20
Pipe	60	15	40	10	-33
Tank	116	29	152	38	+31
Total:	818	176	975	209	+19

estimates of the rigid urethane foam market in construction, according to the five forms in which urethane insulation is most frequently produced, is shown
in Table VI (68). The largest category of urethane
insulation in 1976 will be boardstock, used mainly
for roof and perimeter insulation. Poured-in-place
foam is employed most frequently in factory fabricated
panels and for on-site insulation of cavity walls and
is expected to increase 33 percent over 1975 provided
that the industry can develop metal or other suitable
material-faced urethane foam panels with acceptable
flame spread and smoke evolution (68). Sprayed-on
urethane foam is being used for roof insulation and for
the insulation of perimeters, precast concrete walls
and other irregular surfaces. Large gains in 1976
are predicted for tank insulation -- 31 percent (68),
which is primarily applied by means of spraying-on at
the job site. Pipe insulation is the only category
that has been forecast to decline in 1976, a drop of
33 percent (68), due mainly to the near completion of
the Alaska pipeline.
    The principal reason for the growing acceptance
and popularity of rigid urethane foams is the thermal
conductivity, which is lowest among commercially used
insulating materials, representing significant energy
savings. A comparison of the efficiency of common
tank insulating materials is given in Table VII (69).

Table VII.
COMPARATIVE EFFICIENCY OF COMMON TANK INSULATING MATERIALS

Material	K Factor* or Thermal Conductivity	R Factor*, 1-in. Thicknesses	Thickness in Inches Required for Equivalent Insulating Value
Urethane Foam	0.14	7.1	1.0
Glass Fiber	0.22	4.5	1.6
Styrene Foam	0.28	3.6	2.0
Styrene Board	0.28	3.6	2.0
Mineral Wool	0.30	3.3	2.1
Regranulated Cork	0.30	3.3	2.1
Calcium Silicate	0.35	2.9	2.5
Foam Glass	0.35	2.9	2.5
Asphaltic Paint	10.0	0.1	71.0
Rust Inhibiting Paint	—	Negligible	—

* The lower the K factor and the higher the R, the greater effectiveness as an insulating material. The R factor, which is the reciprocal of the K factor, is a measure of the resistance of a material to transmission of heat and cold.

Table VIII.   BUOYANCY

Material	Density lb per cu ft	Approx. pounds of support in water for each cubic foot
RIGID URETHANE FOAM	2	60
Polystyrene	1.5	60
Balsa Wood	6	56
Cork	7	55
Pine Wood	28	34
Oak	48	14

Other important rigid foam applications are in appliances, transportation, furniture, packaging and marine and flotation. Major usages for rigid foam in appliances include insulation of freezers and refrigerators, and in transportation, insulation of refrigerated railroad cars and trailer trucks. Rigid urethane foam is used for marine salvage and for insulation of ships and old barges in order to extend their lives. A comparison of the buoyancy of rigid urethane foam with other materials is shown in Table VIII (69).

Other growth areas among rigid foams are in the furniture industry (high density foams) and in packaging where in most cases very low density foams (0.5 - 1.0 lb/cu ft) are being used. More recently, rigid urethane foams have been utilized in the manufacture of structural foams (3), and this market promises to be a major factor in the future.

Dr. Irani, Executive Vice President for the Chemicals Group of Olin Corporation, predicted that the production of rigid urethane foam in the next ten years would increase to 1.3 billion pounds (70) with construction being the most important factor in this growth. A composite for the total urethane industry projection by Mobay market researchers through 1980 is depicted in Figure 3 (3).

Sugar or sugar derivative-based polyols have found the widest applications in rigid urethane foams for various industries. However, higher equivalent weight polyols (low hydroxyl numbers) of these sugar or sugar

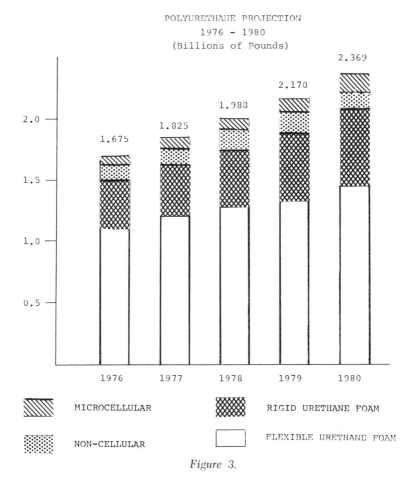

Figure 3.

derivatives may be employed in flexible or semirigid
foams, particularly when blended with conventional
flexible foam polyols (e.g. based on glycerol or trimethylolpropane).   In addition, urethane resins
based on dextrose and maltose polyether polyols have
been found useful as resin binders for foundry use
(71).   Other uses include solid urethane plastics
(72) and foam coatings (73).   In addition, RIM (reaction injection molding) technology could utilize sugar
polyols for the molding of solid plastics, elastoplas-
tics and rigid and semirigid foam products.
Polyether polyols for rigid urethane foams currently
are (August, 1976) priced at 37 - 45 cents/pound while
special, flame-retardant polyols command premium prices

of 70 cents/pound or more.   Polyethers based on sucrose and sorbitol are in the lower price range of the
rigid foam polyols (37 - 40 cents/pound).

Conclusions

Sugar or sugar derivatives are playing important
roles in the manufacture of polyols for urethane polymers, in particular for use in rigid urethane foams.
In view of the anticipated growth of this market
(average growth of 12 percent per year) and the combination of the relatively low costs of sugar-based polyols and excellent physical properties of the resulting
urethane polymers, sugar and sugar derivatives are
expected to enjoy continued future growth.

Abstract

Urethanes have grown at an astounding pace in the
last twenty years in the United States and worldwide.
The total urethane production in the U.S.A. in 1956
amounted only to 8 million lb and reached 1.3 billion
lb in 1975.   The principal components of urethanes
are hydroxyl-terminated polyethers and polyesters -
referred to briefly as polyols- and di- and polyisocyanates.   Flexible and rigid foams make up the bulk of
the urethane market, 1.4 billion lb (1975) and it is
in the foam area that sugars and sugar derivatives
have found their widest acceptance, particularly in
the rigid foam field.   The most commonly used sugar
derivatives are propylene oxide adducts of sucrose,
sorbitol, starch-derived glucosides, dextrose, etc.
Frequently these polyols are modified either by blending with other polyols or employing water, polyhydric
alcohols, amines or aminoalcohols as coinitiators prior
to propoxylation.   This is being done in order to reduce the viscosity of the polyol or to impart certain
desirable properties, such as reduced friability, to
the resulting foams.   A comparison is made of sugar or
sugar derivative-based polyols with other competitive
products as far as price structure and performance
characteristics are concerned.   Some present uses and
future potential of sugars in the urethane foam
market are discussed.

Literature Cited

1.   Upjohn Polymer Chemicals Division "Urethane '76,
      Market Review and Outlook".
2.   Chemical Week, (1976), 118, (15), 32.

3. Mobay Chemical Corp., "Urethane Market Report 1975".
4. Saunders, J.H., and Frisch, K.C., "Polyurethanes" Part I, pp. 34-35., Interscience, New York (1962).
5. Frisch, K.C., "Fundamental Chemistry and Catalysis of Polyurethanes" in "Polyurethane Technology" edited by P.F. Bruins, Interscience-Wiley, New York (1969).
6. Pizzini, L.C., and Patton, J.T., "1,2-Epoxide Polymers" in "Encyclopedia of Polymer Science and Technology" edited by H.F. Mark, N.G. Gaylord and N.M. Bikales, Vol. 6, p. 145, Interscience, New York, (1967).
7. Gee, G., Higginson, W.C.E., and Merrall, G.T., J.Chem. Soc., (1959), 1345.
8. Ishii, Y., and Sakai, S., "1,2-Epoxides" in"Ring-Opening Polymerization", edited by K.C. Frisch and S.L. Reegen, pp. 13-100, Marcel Dekker, Inc., New York, (1969).
9. Price, C.C., "Polyethers" in "Polyethers", Chapter 1, edited by E.J. Vandenberg, ACS Symposium Series 6, Am. Chem. Soc.,Washington D.C.(1975).
10. Lundsted, L.G., and Schmolka, I.R., "The Synthesis and Properties of Block Copolymer Polyol Surfactants" in"Block and Graft Copolymerization", Vol. 2, edited by R.J. Ceresa, John Wiley & Sons, New York (1976).
11. Anderson, A.W., U.S. Pat. 2,902,478 (to Dow Chemical Co.), (1959).
12. Anderson, A.W., U.S. Pat. 2,927,918 (to Dow Chemical Co.) (1960).
13. Patton, J.T., Jr., and Pizzini, L.C., U.S. Pat. 3,454,646 (to Wyandotte Chemicals Corp.)(1969).
14. Davis, P. U.S. Patent 3,251,903 (to Wyandotte Chemicals Corp.) (1966).
15. Davis, P., U.S. patent 3,254,057, (to Wyandotte Chemicals Corp.).
16. Vogt, H.C., and Davis, P., Polymer Eng. and Science, (1971), 11, 312.
17. Pitts, J. J., Fuzesi, S., and Andrews, W.R., J. Cell. Plastics, (1972), 8, (5), 274.
18. Bruson, H.A., Rose, and Rose, J.S., U.S. Pat. 3,244,754 (to Olin Mathieson Chemical Corp.) (1966).
19. Fife, H.R., and Roberts, F.H., U.S. Pat. 2,448,664 (to Carbide and Carbon Chemical Corp.) (1948).
20. Dyer, E., Taylor, H.A., Mason, S.J. and Sampson, J., J. Am. Chem. Soc.,(1949), 71, 4106.

21. Wissman, H.G., Rand, L., and Frisch, K.C. J.Appl.
     Polymer Sci., (1964), 8, 2971.
22. Rand, L., Thir, B., Reegen, S.L., and Frisch, K.C.
     J. Appl. Polymer Sci., (1965), 9, 1787.
23. Knodel, L. R., U.S. Pat. 3,865,806 (to Dow Chem-
     ical Co.) (1975).
24. Nawata, T.,Kresta, J.E., and Frisch, K.C.,J. Cell.
     Plastics, (1975), 11, (5), 267.
25. Private Communication.
26. Wismer, M. and Foote, J.F., U.S. Pat. 3,085,085
     (to PPG Industries) (1963).
27. Patton, J.T., Jr., Schulz, W.F., U.S. Pat. 3,346,
     557 (to Wyandotte Chemicals Corp.) (1967).
28. Ulyatt, J.M., U.S. Pat. 3,357,970 (to Chas. Pfizer
     & Co.) (1967).
29. Brandner, J.D., U.S. Pat. 3,359,217 (to Atlas
     Chemical Industries) (1967).
30. Degginger, E.R., and Booth, R.E., U.S. Pat. 3,442,
     888 (to Allied Chemical Corp.) (1969).
31. Katsunuma, H., and Ishizuka, Y., Jap. Pat. 71-278-
     15 (to Asahi Electro-Chemical Co.) (1971).
32. Kaiser, D.W., and Fuzesi, S., U.S. Pat. 3,167,538
     (to Olin Mathieson Chemical Corp.) (1965).
33. Booth, R.E., U.S. Pat. 3,369,014 (to Allied Chemi-
     cal Corp.) (1968).
34. Aitken, R.R., and Marklow, R.J., U.S. Pat. 3,446,
     848(to Imperial Chemical Industries Ltd.)(1969).
35. Booth, R.E., and Degginger, E.R., U.S. Pat. 3,424,
     700 (to Allied Chemical Corp.) (1969).
36. Fijal, W.R., U.S. Pat. 3,640,997(to BASF-Wyandotte
     Corp.) (1972).
37. Wismer, M., Lebras, L.R., Peffer, J.R., and
     Foote, J.F., U.S. Pat. 3,153,002 (to Pittsburgh
     Plate Glass Co.) (1964).
38. Wismer, M., Lebras, L.R., Peffer, J.R., and Foote,
     J.F., U.S. Pat. 3,222,357 (to Pittsburgh Plate
     Glass Co.) (1965).
39. Wismer, M., Lebras, L.R., and Foote, J.F., U.S.
     Pat. 3,265,641 (to Pittsburgh Plate Glass Co.)
     (1966).
40. Wilson, J.E., Truax, H.M., and Dunn, M.S.,J. Appl.
     Polymer Sci. (1960), 3, (9), 343.
41. Gibbons, J.P., and Wondolowski, L., U.S. Pat.
     3,816, 395 (to CPC International) (1974).
42. Hostettler, F., Barnes,R.K., and McLaughlin, R.W.,
     U.S. Pat. 3,073,788 (to Union Carbide Corp.)
     (1963).
43. Yotsuzuka, M., and Fujima, S., Jap. Pat. 72-41437
     (to Takeda Chemical Industries, Ltd.) (1972).

44. Winquist, A.D., Jr., and Theiling, L.F., Jr., U.S.
    Pat. 3,317, 508 (to Union Carbide Corp.)(1964).
45. Fuzesi, S., and Klahs, L.J., U.S. Pat. 3,402,170
    (to Olin Mathieson Chemical Corp.) (1968).
46. Othey, F.H., and Mehltretter, C.L., U.S. Pat.
    3,165,508 (to United States of America as
    represented by the Secretary of Agriculture)
    (1965).
47. Moss, P.H., and Cuscurida, M., U.S. Pat. 3,721,665
    (to Jefferson Chemical Co.) (1973).
48. Othey, F.H., Bennett, F.L., Lagoren, B.L., and
    Mehltretter, C.L., Ind. Eng. Chem. Prod. Res.
    Develop. (1965), 4, (224), 228.
49. Leitheiser, R.H., Impola, C.N., Reid, R.J., and
    Othey, F.H., Ind. Eng. Chem. Prod. Res.
    Develop. (1966), 5, 276.
50. Othey, F.H., Westhoff, R.P., and Mehltretter, C.L.,
    J.Cell. Plastics, (1972), 8, (3), 156.
51. Wilham, C.A., Othey, F.H., Mehltretter, C.L.,
    and Russell, C.L., Ind. Eng. Chem. Prod. Res.
    Develop., (1975), 14, 189.
52. Heckles, J.S., and Quinn, E.J., U.S. Pat. 3,764,
    570 (to Armstrong Cork Co.) (1973).
53. Anderson, J.J., U.S. Pat. 3,251,785 (to Socony
    Mobil Oil Co.) (1966).
54. Friedman, L., U.S. Pat. 3,261,814 (to Union Car-
    bide Corp.) 1966).
55. Lutz, M.R., U.S. Pat. 3,251,828 (to FMC Corp.)
    (1966).
56. Lankro Chemicals, Brit. Pat. 1,105,953 (1968).
57. Patton, J.T., Jr., and Hartman, R.J., U.S. Pat.
    3,530,205 (to Wyandotte Chemicals Corp.)(1970).
58. Quinn, E.J., Ind. Eng. Chem. Prod. Res. Develop.,
    (1970), 9, (1), 48.
59. Birum, G.H., U.S. Pat. 3,317,510 (to Monsanto Co.)
    (1967).
60. Pape, P.G., Sanger, J.E., and Nametz, R.C., J.Cell
    Plastics, (1968), 4, (11), 438.
61. Austin, A.L., Pizzini, L.C., and Levis, Jr., W.W.,
    U.S. Pat. 3,639,541 (to BASF Wyandotte Corp.)
    (1972).
62. Papa, A.J., "Flame-Retardant Polyurethanes" in
    "Flame Retardancy of Polymeric Materials",
    Vol. 3, edited by W.C. Kuryla and A.J. Papa,
    Marcel Dekker, New York (1975).
63. Frisch, K.C., and Reegen, S.L., "Relationship
    Between Chemical Structure and Flammability
    Resistance of Polyurethanes" in "Flame-Retard-
    ant Polymeric Materials" edited by M. Lewin,
    S.M. Atlas and E.M. Pearce, Plenum Press,

New York (1975).
64. Darr, W.C., Gemeinhardt, P.G., and Saunders, J.H.
    Ind. Eng. Chem. Prod. Res. Develop. (1963),
    2, 194.
65. Frisch, K.C., J. Cell. Plastics, (1965), 1, (2),
    321.
66. Reegen, S.L., and Frisch, K.C., J. Polymer Sci.
    (1967), 16, 2733.
67. Saunders, J.H., and Frisch, K.C., "Polyurethanes"
    Part II, Interscience, New York (1964).
68. Urethane Plastics and Products(1976), 6, (6), 1.
69. "U.S. Foamed Plastics Markets and Directory 1976",
    Technomic Publishing Co., Westport, Conn.
70. Urethane Plastics and Products, (1976), 6, (5), 3.
71. Molotsky, H.M., U.S. Pat. 3,743,621 (to CPC Inter-
    national) (1973).
72. Deanin, R.D., Ellis, E.J., Briere, T.A., and Dunn,
    M.R., SPE Journal, (1972), 28, (4), 56.
73. Karashi, A., and Kitakawa, T., Jap. Pat. 71-01710
    (to Takeda Chemical Industries) (1971).

Biographic Notes

        Professor Kurt C. Frisch, Ph.D., Prof. of Polymer
Eng. and Chem. and Dir. of Polymer Inst.   Educated
at Univ. of Vienna, Univ. of Brussels and Columbia
Univ.   Industrial experience at General Electric Co.,
E.F. Houghton & Co. and Wyandotte Chemical Corp.
Joined Univ. of Detroit first in Chem. Eng. Dept. and
then at the Polymer Inst.   Some 100 publications and
books and 35 patents in polymers. PoPolymer Inst., Coll.
of Eng., Univ. of Detroit, 4001 West McNichols Rd.,
Detroit, Michigan, 48221 U.S.A.

Sucrose and Modified Sucrose Polyols in Rigid Urethane Foam

ALLAN R. MEATH and L. D. BOOTH

Dow Chemical Co., Bldg. B-4810, Freeport, Tex. 77566

Sucrose is used widely as an initiator for polyols to make rigid urethane foam. The primary reasons for the use of sugar are, (a) it is readily available from a number of sources, (b) laboratory evaluations have shown that sugar derived from either sugar beets or sugar cane is acceptable, (c) consistent quality and purity of sugar make it an ideal chemical starting material, (d) sugar with eight reactive hydroxyl groups has the functionality necessary to obtain the high degree of crosslinking needed to produce a rigid urethane foam, and (e) price.

A polyol, useable for rigid urethane foam, is prepared by the addition of an alkylene oxide to the sucrose molecule. Alkylene oxides used are:

> Ethylene Oxide,
> Propylene Oxide, and
> Butylene Oxide.

One or more molecules of alkylene oxide is added to each of the eight reactive hydroxyl groups on the sucrose molecule. The resulting product is a liquid polyol with eight reactive hydroxyl groups.

The choice of which alkylene oxide is used has several effects. The choice of oxides effects the reaction rate of the alkylene oxide addition to the sucrose molecule. It also has an effect on the physical characteristics of the resulting polyol and plays a major role on the physical properties of the final rigid urethane foam. Table I shows the effects of using ethylene oxide, propylene oxide, and mixtures of the two oxides. Butylene oxide offers an advantage over the other oxides in the area of hydrolytic stability. However, butylene oxide is not widely used because it commands a high price and cannot be justi-

fied on a price for performance basis.   As can be seen
from Table I, ethylene oxide offers advantages in the
following areas: a faster reaction rate for the oxide
addition to sucrose; a resulting polyol with a lower
viscosity; and a polyol containing primary hydroxyl
groups which increase the polyol's reactivity with an
isocyanate.   However, ethylene oxide has a major disadvantage in that an ethoxylated polyol produces a
rigid urethane foam with poor hydrolytic stability.
Since the major applications for rigid urethane foams
require good hydrolytic stability, propylene oxide is
the most widely used alkylene oxide.   In the final
analysis, the type or ratio of alkylene oxide used is
dependent on the intended end uses.

Table I.

Effect of Different Oxides on Polyol
Preparation and End Properties

Ethylene Oxide
    Faster oxide addition rates
    Lower polyol viscosity
    Primary hydroxyl groups
    Faster urethane reaction
    Poor humidity aging properties
Propylene Oxide
    Secondary hydroxyl groups
    Slower urethane reaction
    Better humidity aging stability
Mixture of Ethylene and Propylene Oxide (4)
    Faster oxide addition
    Primary and secondary hydroxyls
    Lower polyol viscosity
    Good humidity aging properties

        Also affecting the sucrose initiated polyol is the
amount of alkylene oxide added.   Table II shows, gra-
phically, that, as you increase the amount of oxide
added to the sucrose molecule, the lower will be the
resulting polyol viscosity and the less rigid the foam.
With the addition of only one alkylene oxide per hydroxyl group on the sucrose molecule, there are some
applications where the resulting foam is too rigid or
friable.
        The alkylation of the sucrose molecule can be
accomplished using a high pressure and temperature
reaction.   If a coinitiator is used, the reaction can
be carried out at low pressure.   The coinitiator is
a liquid material having "labile" hydrogens, (1) reac-

tive enough to react with propylene or ethylene oxide
and in which the sugar is soluble.

Table II.
Oxide Level vs Polyol Properties

300	400	500	600 Hydroxyl
	Usable viscosity	High viscosity	Number
	Reduced polyol cost	Foam very rigid	
	Useful for:	Foam useful for:	
	Billets (boardstock)	Spray	
	Pour-In-Place	Thin section –	
	High density	Pour-In-Place	
			Oxide/OH
	2 PO/OH		1 PO/OH

   Table III shows the raw material charge and reaction conditions for a high pressure, polyol preparation
reaction where sucrose was used as the sole initiator.
While the entire charge of sucrose, alkylene oxide,
and catalyst may be mixed before the reaction is initiated, this may result in undesirably vigorous reaction
and poor temperature control, especially when ethylene
oxide is being used.   A preferred procedure comprises
mixing the sucrose, the catalyst and a small portion of
the alkylene oxide, heating the mixture to reaction
temperature and then, when the reaction has begun,
feeding in the remaining oxide at about the rate at
which it reacts, thus permitting a steady rate of
reaction and effective temperature control (2).

Table III. Example of High Pressure Process (2)

Reactor - Pressure Autoclave

Charge - Sucrose 2400g
          Propylene Oxide 3600g
          Trialkylamine 21g
            or
          Metal Hydroxide

Reaction Conditions
   Temperature
      105 - 110°C for 6 hr
      110 - 115°C for 4 hr
      105°C   for 2 hr

   Pressure
      Maximum - 107 psig
      Final   -   30 psig

Materials with reactive hydrogens which are used as coinitiators with sucrose to make rigid polyol include:

Water,
Trimethylolpropane,
Glycerol,
Ethylenediamine,
Diethylenetriamine,
Propylene Glycol, and
Dipropylene Glycol.

An example of a raw material charge and reaction conditions for a low pressure process preparation of a sucrose-glycerol, coinitiated polyol is given in Table IV. The sucrose, glycerol, and catalyst were pre-mixed until the sugar was dissolved and then the mixture was preheated to 130°C. The propylene oxide was added over a 12 hr period. During this period, the pressure maintained by the propylene oxide was from 30-40 psig and the temperature maintained at 125 to 135°C. Upon completion of the propylene oxide addition, the reaction mixture was digested for 2 hr at 130°C (3).

Table IV. Example of Low Pressure Process (3)

Reactor   -   Low Pressure Vessel

Charge   -
    Sucrose . . . . 17.1 lb
    Glycerine . . . . 11.5 lb
    Trialkylamine . . . . 105 g
    Propylene Oxide 45   lb

Reactor Conditions
    Temperature
        125 - 135°C for duration of the reaction
    Pressure
        30 - 40 psig

The sucrose, glycerol, and amine catalyst were pre-mixed until the sugar was dissolved as the mixture was preheated to 130°C. The propylene was added over a 12-hr period.

Advantages of using a coinitiator with sucrose include:

Increased speed of alkylation reaction,

Decreased polyol functionality,
Decreased polyol viscosity,
Improved compatability with polymeric
  isocyanate and fluorocarbon blowing
  agent,
Ability to use low pressure processing, and
Elimination of necessity of handling and
  blending high viscosity sucrose initiated polyol.

Actually, there are specific advantages to using
each type of coinitiator shown above.   For example:

H_2O

Faster alkylation reaction,
Diol to polyol introduction,
Improved compatability,
Decreased polyol viscosity, and
No change in polyol reactivity.

Amines -

Increased speed of alkylation process,
Low pressure processing,
Decreased polyol functionality,
Decreased polyol viscosity,
Improved compatability,
Increased polyol reactivity,
Internal catalyst, and
Increased raw material cost.

Glycerol -

Increased speed of alkylation,
Decreased polyol functionality,
Decreased polyol viscosity,
No change in polyol reactivity, and
Increased polyol raw material cost.

Polyols based on sucrose or sucrose and a coinitiator are used primarily to make rigid urethane foams
for a wide variety of end-uses.   Following are some
applications listed by application method.

Spray Foam
  Insulation
    Building (roofs)
    Reaction and storage vessels

Pour-In-Place
   Insulation
      Refrigeration
      Portable Coolers
      Pipe Coverings
      Sandwich Panels
   Buoyancy
      Boats
      Barge Repair
   Packaging
   Furniture - Wood Replacement
Billets
   Insulation
      Pipes and Vessels
      Cold Storage Rooms
      Commercial Buildings
      In conclusion, sucrose is a very good candidate
as the base initiator for preparation of rigid urethane
foam.   Its bicyclic structure and eight reactive sites
provide the urethane polymer with good thermal and
dimensional stability.
      Various applications have different physical pro-
perty requirements.   Sucrose with its high functionality allows one to blend it with a variety of coinitiators to meet the various requirements.

Abstract

      Sucrose-initiated polyols for use in rigid ure-
thane foams can be made commercially by methods described as high pressure and low pressure processes.
The advantages and disadvantages of these two processes
are discussed.   In the low pressure process it is necessary to dissolve or disperse the sugar in a coinitiator.   Commonly used are water, trimetholpropane, glycerol, ethylenediamine, and diethylenetriamine.
The choice of coinitiator changes the resulting polyol
and the final rigid urethane foam's properties.
      The choice of oxide and the amount of oxide added
to the sucrose and coinitiator will vary the average
molecular weight and the ratio of primary and secondary
hydroxyl groups.   These changes affect the rate of reactivity of a polyol with an isocyanate and the foam
end-use physical properties.

Literature cited
1.   United States Patent 2,990,376
2.   United States Patent 2,902,478
3.   United States Patent 2,990,376
4.   United States Patent 3,865,806

Biographic Notes

Allan R. Meath, M.Sc., M.B.A., Group Leader in Alkylene Oxide Derivatives.   Educated at North Dakota Univ. and Central Michigan Univ.   Joined the Dow Chemical Co., in 1955, specializing in latexes, epoxy resins and urethane chemicals.   Dow Chemical Co., Building B-4810, Freeport, Texas 77566 U.S.A.

19

Sucrose-Based Rigid Urethanes in Furniture Applications

STEPHEN FUZESI

Olin Corp., 275 Winchester Ave., New Haven, Conn. 06504

The furniture industry is moving from conventional materials, mainly wood, to plastic foams to effect cost savings in labor and materials. By 1985,industry estimates suggest, that plastics will account for 35 percent of furniture costs and may amount to $3 billion(1) in sales volume. (Figure 1).

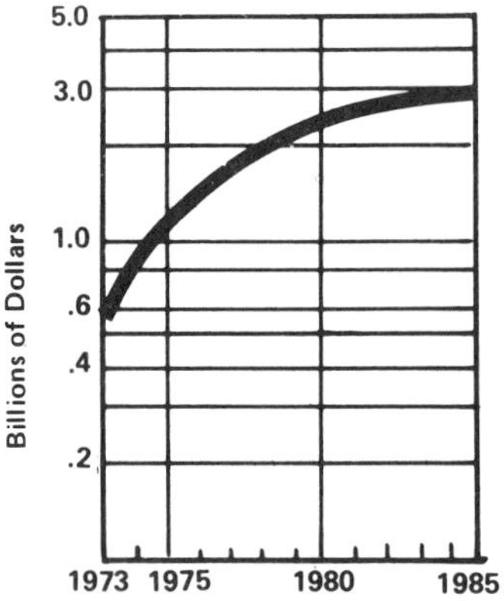

Figure 1. Value of plastics used in furniture manufacture (Source: U.S. Department of Agriculture, Industrial Outlook, 1974; SRI and industry estimates)

The market for moldable,high density,urethane foam **also** is well established, and is growing rapidly. More than 1.0 billion pounds of plastics were used in the furniture industry in 1974. Fifty million pounds, 5 percent of the total, was accounted for by rigid urethane foams (Table I). This figure is expected to grow to 150 million pounds per year by 1980.

Table I.   Rigid Urethane Foam Markets (3)
           (millions of pounds)

	1970	1972	1973	1974	1975	1980
Appliances	42	60	75	90	96	120
Construction	75	130	155	160	160	290
Marine	8	10	13	12	20	30
Vehicles	37	40	50	55	60	80
Furniture	23	29	40	50	70	150
Miscellaneous	6	6	8	5	10	15
Total	191	275	341	372	416	685

The structure and the properties of high density, rigid urethane foams can be controlled (a) by chemical formulations and (b) by variation of the processing parameters. The technology of molding rigid urethanes has advanced rapidly during recent years. Newly developed molding techniques now make possible;

a. The production of finished parts with solid skins and micro-cellular interiors in a single operation from the same raw materials. This type of urethane foam is called self-skinned or integral skinned structural foam.

b. The production of finished parts with homogeneously uniform cell structures, possessing very thin skin or with practically no skin at all. This type of urethane foam is called conventional, high density, molded, rigid urethane wood foams.

The chemical components (Table II) in integral skin, rigid urethane foams are similar to those used in conventional, high density, rigid urethane wood foams, and include: polymeric isocyanates; catalysts; blowing agents; surfactants; and polyols.

Polymeric isocyanates like PAPI-135 (Upjohn Co.), Mondur MR (Mobay Chemical) and Rubinate-M (Rubicon Chemical) are the most widely used.

Catalysts used are both amines and metal salts, mainly tin based, and sometimes in combinations. The amine-type catalysts initiate the polymer formation and promote mold fill. The tin-type catalysts promote the final gelling, complete the cure and control the demolding time.

The surfactants regulate the interfacial tension in the foaming system and control uniform cell formation. High density, rigid foams do not require sophisticated surfactants, most commercial products perform well (2).

Both water and fluorocarbons are used as blowing agents, sometimes in combination. Water is used mainly in preparation of the conventional, high density,

Table II

Major Components in High Density Rigid Urethane Wood Foams

1. Isocyanate-Polymeric   Polymer Formation
   PAPI-135[1]   Polyurethane
   Rubinate-M[2]   Polyurea
   Mondur MR[3]   Isocyanurate

2. Catalyst
   (A) Amine Catalyst
       Dabco[4]   Initiate the polymer formation.
       Polycat-8[5]   Blow and promote mold fill.
       TMBDA[6]
   (B) Organotin Catalyst
       DBTDL[7]
       Stannous Octoate[8]   Promote the final gelling, Complete the
       UL-6(blocked)[9]   cure.   Control the demolding time.
       UL-24(blocked)[10]

3. Surfactant
   Dow DC-193[11]   Regulate the interfacial tension
   U.C.-5420[12]   in the foaming system.   Control
   U.C.-5430[13]   the uniform cells formation.

4. Water (H_2O)
   Used in low level only   Polyurea and CO_2(↑)-Improve the
                                           rigidity of the foam.   Reduce the
                                           skin density.

5. Fluorocarbon-blowing agent
   R-11B[14]   Blow the foam.   Control the skin
                                           thickness and the foam density.

6. Polyols   Polyurethane formation.
                                           Most diversified and most critical.

---

1. Upjohn	4. Air Product	7. M & T	10. Argus	13. Union Carbide
2. Rubicon	5. Abbott	8. M & T	11. Dow Corning	14. DuPont
3. Mobay	6. Witco	9. Argus	12. Union Carbide	

Table III

Selection of the Polyols for Wood Foam Applications

	Optimum	
Hydroxyl numbers	330 - 430	Crosslink density
Functionalities	4 - 5	Strength
Cyclic backbones	Carbohydrate-based	Structure
		Hardness, Brittleness
Viscosity	<3,000 cps @ 25°C	Handling
Reactivity	Well balanced built	Flow, Mixing
	in reactivity	Cream, Gel
		Exotherm-split
		Demolding
		Economy
Other	Amines built in the	Compatibility with poly-
	polyol backbone	meric isocyanate in mixing
		and in the molded parts

wood foams. Fluorocarbon is used mainly in the formulation of integral skin, structural foams.
The choice of the proper polyol is most critical.
While the specific criteria for performance of a polyol
for furniture application may vary from customer to
customer, the key requirements are similar (Table III).
These include: low polyol viscosity - 3,000 cps max;
good polyol compatibility with polymeric isocyanate,
water and fluorocarbons; excellent flow in the mold
cavity; rapid demolding time; and the resultant foam
must possess acceptable strength and good dimentional
stability.
The viscosity and the reactivity of the polyol
are the most important with respect to handling, mixing, flow and demolding of the foam. The hydroxyl
number, functionality and the backbone structure of the
polyol are the most important in controlling the cell
and skin structure, as well as the strength and hardness of the foams.
Our work to date, indicates that sucrose amine-
based polyols offer the most versatility in obtaining
all of these criteria. Sucrose is a non-reducing disaccharide (Figure 2). It is a bicyclic compound with
eight functionalities. It is stable in a basic condition and readily reacts with epoxides in polyether formation. Sucrose-based polyols are compatible with
other nonsugar-based polyols. Sucrose is available on
the market by competitive suppliers. Prices are reasonable, although sometimes they fluctuate.

Sucrose

(Glucose portion)   (Fructose portion)

Nonreducing disacchaoride.
Cyclic compound with high (8) functionalities.
Stable in basic condition.
Readily react with oxides in polyether formation.
Sucrose based polyols are compatible with other non-sugar
  based polyols.
Available on the market by competitive suppliers.
Price reasonable although fluctuate.

Figure 2

The most suitable polyols are prepared by a coinitiation technique (Table IV) whereby optimal functionalities, hydroxyl numbers, viscosities and reactivities
can be achieved.

Table IV.

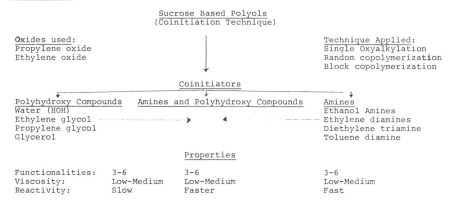

Sucrose Based Polyols
(Coinitiation Technique)

Oxides used: Technique Applied:
Propylene oxide   Single Oxyalkylation
Ethylene oxide   Random copolymerization
                                                        Block copolymerization

Coinitiators

Polyhydroxy Compounds	Amines and Polyhydroxy Compounds	Amines
Water (HOH)		Ethanol Amines
Ethylene glycol		Ethylene diamines
Propylene glycol		Diethylene triamine
Glycerol		Toluene diamine

Properties

Functionalities:	3-6	3-6	3-6
Viscosity:	Low-Medium	Low-Medium	Low-Medium
Reactivity:	Slow	Faster	Fast

Laboratory data, derived from comparative evaluations, have shown that properly designed, sucrose amine
based polyols are suited ideally for the rapidly growing, furniture market.   Olin has developed such a polyol, which we believe is capable of meeting and exceeding the customer's performance criteria for high density rigid foams, mainly in the furniture applications.
Poly-G 71-357 polyether polyol is a new product developed for use in molded, high density,rigid foam applications.   It is a sucrose-amine-propylene oxide-based
polyol.   The chemical and physical properties of this

Table V.

Shipping Specifications and Physical Properties
Data for Poly-G 71-357

Shipping Specifications		Physical Properties	
Hydroxy No.,(mg.KOH/g	350+10	Viscosity, (cps, 77°F)	2,000
Color (Gardner-max.)	12	Specific Gravity	1.08
Water, (% by wt.-max.)	0.08	Lbs per gallon	9
pH, (in 10/6 Isopropanol/	9.5+1	Flash Point, *(°C)	180
water		Pour Point, (°C)	-12

*It must be realized that flame properties of this or any other
raw material are not intended to reflect the fire hazards presented
by any cellular or foamed plastic product containing this raw material.

polyol are shown in Table V.   The hydroxyl number of
the polyol is 350, the viscosity 2,000 cps at 25°C.
The pH is 9.5 + 1, a good indication of satisfactory
reactivity.
   The lower hydroxyl number reduces formulation cost,
by reducing the required isocyanate, but it does not
compromise important physical properties due to its
higher functionality from the sucrose-based cyclic
backbone built into the polyol.   The quantity of
catalyst required to promote the desired reaction profile is less than would be required with polyols with
no amine in the backbone.
   The temperature-viscosity curves, including the
effect of fluorocarbon-11 blowing agent, are shown in
Figures 3 and 4.   This low viscosity polyol provides
easier handling, good mixing, and excellent flow.
Laboratory data, derived from comparative evaluations,
have shown this polyol to be inherently more compatible
than the other polyols tested with polymeric isocyanate, water and fluorocarbons.
   A typical foam formulation common in conventional,
high density, urethane wood foam, and the resultant
foam properties are shown in Tables VI and VII, respectively.
   The compressive strength of the foam is density
dependent, and may vary within a single sample depending on: (1) thickness of the specimen; (2) skin thickness; and (3) cell structure of the foam.   Compressive
strengths of 337 psi parallel to rise, and 311 psi per-
pendicular to rise on a foam of 11.5 pcf core density
are excellent values.

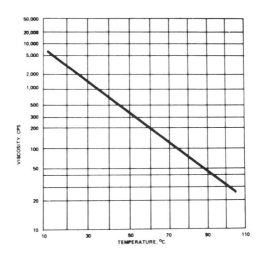

Figure 3. Viscosity vs. temperature, Poly-G
71-357

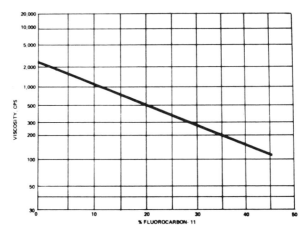

Figure 4.   Viscosity vs.   fluorocarbon-11   concentration,
              Poly-G 71-357, at 25°C

Nail and screw holding characterisitics are important in many furniture applications. The screw-
holding power of the foams, of course, varies with
the density of the foams and the screw diameter. The
higher the density of the foam, the higher its screw
holding power. It should be noted that, because the
urethane foam does not have grain, there is much less
chance of splitting taking place.
    Besides the excellent compressive strength, the

Table VI.

Handmix Formulation for Poly-G 71-357 (12 PCF Foam)

"B" Component	% by Weight	pbw
Poly-G 71-357	97.08	100.00
DC-193[1]	1.46	1.5
Dabco 33LV[2]	0.97	1.0
Water	0.49	0.5
	100.00	103.0

"A" Component		
PAPI-135[3]	100.0	94.0

Processing Data

Mix Ratio A/B	0.91/1	
Reactivity Data:		
Cream (Sec.)	40	(1) Dow Corning
Rise (sec.)	80	(2) Air Products
Tack free (sec)	90	(3) Upjohn

Table VII.

Physical Properties of Poly-G 71-357 Hand Mix Foam

Formulation Test No. 79818 Density, core, pcf	11.5
Compressive Strength, psi, at yield	
Parallel to rise	337
Perpendicular to rise	311
Screw Holding, lbs. (1 1/2" Screws)	70
Impact, lzod, (in. lb./in.)	5.6
Humid Age, 158°F/100% R.H., % ΔV (max.), 28 days	<1
Dry Heat Age, 200°F, ambient R.H., % ΔV (max.), 28 days	<1

screw holding (70 lbs) and the impact (Izod 5.6 in. lb/in) values are clear indications of the excellent strength of these 11.5 pcf foams. Of course, many of the physical properties depend on the size, shape, density, thickness and cell structure of the test specimens, so that most of the data published should be considered average values.

Conclusion

A wide variety of decorative as well as structural parts now are being produced from integral skin and from conventional, high density, rigid urethane foams. Complete case goods, with one piece molded drawers, tops, bases and sides with integral fittings can be molded. Frames, shells and free-foam pieces can be made and stapled, screwed, glued and sawed. The material duplicates in detail the mold surface that is used, and textured, wood grain, or highly smooth surfaces may be produced, depending on design considerations.

The application possibilities for the high density rigid urethane wood foams mentioned here are just a few. The list of industries turning to this new plastic foam is growing daily. In the automotive area, hoods, fenders and other trim parts are possible.

We believe that the raw material suppliers are well prepared for this market and the sucrose based polyols will play an important role in this area.

Acknowledgments

The author would like to acknowledge the contribu-
tions of Messrs. J.J. Cimerol, J.S. Sedlak and R.J.
Raynor for the development of the foam formulations.
Physical test data on the foams were obtained within
Olin Urethane Research Department under the direction
of D.R. Shine.

Abstract

The furniture industry is moving from conventional
materials (mainly wood) to plastic foams to effect
cost savings in labor and materials.   Industry estimates suggest that by 1985 plastic will account for
35% of furniture material costs at a dollar value of
3 billion.   Twenty-five thousand metric tons or 5% of
the total plastics consumed by furniture in 1974 was
accounted for by rigid urethane foams.   Estimates
indicate that this figure may double by 1980.   The
chemical components in structural,rigid urethane foams
are similar to those used in conventional rigid foams;
poly-isocyanates, catalysts, blowing agents, surfactants and polyols.   The choice of the proper polyol is
most critical.   Sucrose amine-based polyols offer the
most versatility in obtaining the desired results.
The most suitable polyols are prepared via a coinitiator technique whereby optimal functionality (4-5),
hydroxyl number (330-430), use viscosity (2,500 cps
max) and reactivity are achieved.   Comparative laboratory evaluations have shown that sucrose amine-based
polyols are inherently more compatible with polymeric
isocyanates than other polyols.   They have also
demonstrated excellent flow and demolding characteristics and produce hard but not brittle structural foams.

Literature Cited

1.   Stamford Research Inst., "Long Range Planning
       Service Report", "New Plastic Processing
       Technologies", 1975.
2.   Dunnous, J., "Formulation and Processing Tech-
       nique of high-Density Molded Structural Rigid
       Urethane Foam", Journal of Cellular Plastics,
       (1975), 11, (6).
3.   "U.S. Foamed Plastics Markets and Directory",1975,
       Technomic Publishing Co., Inc., Westport,
       Connecticut, USA.

Biographic Notes

Stephen Fuzesi, Ph.D., Res. Assoc. in Urethane Res. Educated at the Univ. of Dayton, St. John's Univ. (New York) and Peter Parmany Univ. (Budapest). Joined Olin in 1960, working on polyethers in flexible and rigid foams. Rigid Urethanes R & D, Olin Research, 275 Winchester Ave., New Haven, Connecticut 06504 U.S.A.

20

Chemicals by Fermentation—An Overview

ROGER WILLIAMS, JR.

Roger Williams Technical and Economic Services Inc., Princeton, N.J. 08540

Fermentation certainly is the oldest unit process in organic chemistry. When fermentation became commercial is lost in the mists of time, but many of us think of the heyday of fermentation as occurring during World War II, when ethanol was made for butadiene manufacture. Thereafter fermentation was replaced by petrochemical sources.

That was true of ethanol, 'industrial alcohol', as the first chart (Figure 1) shows. Note that this is in dollars, not pounds; and that all the data have been converted to 1975 dollars to wipe out the effects of inflation. One thing that may surprise some, is that the ethanol business, in real dollar terms, really has not grown in the 50 years from 1925 to 1975.

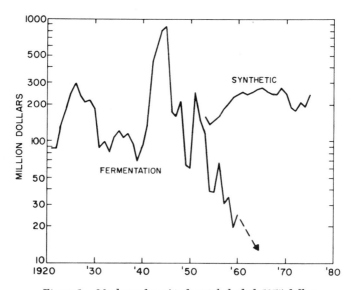

Figure 1. Market value of industrial alcohol, 1975 dollars

274

Of course the other part of the ethanol business, alcoholic beverages, supposedly was a zero business in 1925. It was not, but there are no statistics available on moonshine. Today the liquor business is $1.9 billion, and beer is a hefty $4.8 billion. In total, alcoholic beverages constitute a $7.8 billion business based entirely on fermentation.

Production of butanol-acetone-ethanol jointly by fermentation is another old process which fell by the wayside to petrochemistry. It may be coming back, as will be discussed by Dr. Frank Wynn Hayes, later. At its peak, in the United States, the output probably hit $100 million per year, in 1975 dollars.

The whole antibiotics business has come into being since 1925. Figure 2 shows this business since 1950, again in 1975 dollars. The crash in prices in 1965 is very evident. The figures shown include all the antibiotics. Although some are synthetic, they are in a minority. Thus, antibiotics comprise another three-quarters of a billion dollar business based on fermentation, that did not exist 50 years ago.

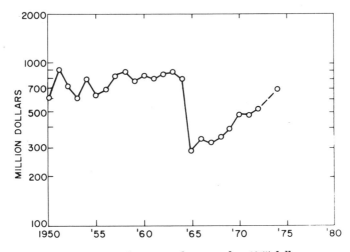

Figure 2. Antibiotics production value, 1975 dollars

Let us consider citric acid, and fumaric, lactic, itaconic, and the other acids that are made by fermentation. Today, perhaps, they are a $100 million business that was non-existent 50 years ago. To this one can add amino acids, although United States production is not dominated by fermentation, and vitamins, and other similar products. In addition, there are other

food products made by fermentation such as vinegar, soy
sauce and yogurt which alone has grown to approximately a $250 million business.

Also, there is the enzyme business, which has
grown to $50 million or so. It may have peaked higher
than that in the enzyme detergent days, but it is
steadily growing. The enzymes for wet corn milling
and high fructose corn syrup production probably have
the highest current growth rate.

In total, then, there is a fermentation business in
the United States today grossing some $8.5 - 9 billion.
That is a very sizable business and it is growing.

One could add to the current fermentation business,
in the United States, all the secondary sewage and waste
treatment plants. Although one cannot put a value on
the product, as one can for the other commercial fermentations, there is a value, and a high one.

Between now and 1985, some 4,000 or more communi-
ties in the United States will be building new sewage
disposal plants with secondary biological treatment,
including those who will have to up-grade primary
plants to secondary plants. To this there is added
industrial waste treatment business.

No attempt has been made to put a dollar value on
just the construction involved. The sludge produced
will be tremendous and, at least, has a fuel value, even
though the whole operation is a net expense to the tax-
payer. The value in human health terms of this fermentation is incalculable.

There does exist a healthy, thriving fermentation
business in the United States today. Can one guess
where it is going? At least there can be outlined
some of the prospects, other than growth of all the
above.

Production of methane from feed lot wastes by
anaerobic fermentation is just beginning. This is one
waste material which already is concentrated geographi-
cally. The first demonstration plant for making
methane from urban refuse, at Pompano Beach, Florida,
is due to start up this year. How successful these
projects will be is still an open question. The main
factor in their favor basically is a negative substrate
cost.

Another project for bioproduction of methane involves growing of kelp in huge beds off the California
coast. That one would seem a bit further off in time.

Again, on the energy scene one is beginning to see
ethanol as a gasoline additive or supplement. In Nebraska gasoline is being promoted, but it is in Brazil that the production of alcohol for motor vehicle

use is a major undertaking. By the mid-1980's they
hope to get production up to some 2 billion gallons per
year, or almost 150,000 barrels per day. They are
projecting fermentation units of up to 2,000 bbls/day
capacity. While some of this will be based on traditional by-product sources such as molasses, Brazil will
also use whole cane juices, and is planning substantial
cultivation of casava, purely for ethanol production.

Single cell protein production (see Wells, Chapter 22) could be big, but more probably overseas rather
than in the United States. There is one point about
SCP that might be mentioned here. For every 100,000
ton per year SCP plant that goes in overseas, the U.S.
loses a potential for soybean or soybean meal export.
Without agricultural exports the U.S. would be hard put
to import almost half its crude oil plus refined products needs. The balance of payments would deteriorate
rapidly. Each 100,000 ton SCP plant costs the U.S.
roughly 1,000,000 barrels of crude in foreign exchange
balance. On the other hand, SCP plants conversely aid
the foreign nations' balance of payments.

A "sleeper" would be the production of "synthetic"
fats and oils from the same sort of substrates used for
SCP. There is much discussion about a world protein
shortage. Actually, the major problem is a calorie
shortage. In many nations, proteins are being used
by the human body for their caloric content, simply
because the human is calorie short. The net result
physically may seem like a protein shortage.

Figure 3 shows a recent Organization for Economic
Cooperation and Development (OECD) chart of calorie
availability. The FAO says the "average" human needs
2,400 calories per day. Keep in mind that, if a na-
tion, or continent, averages 2,400, then half of the
population is starving. Dr. M.G. Krishna,of India's
petroprotein research,estimates that the average figure
should be at least 10% above the minimum requirement to
have 80-90% of the people achieve the bare minimum.
That is the dashed line we have added across the OECD
chart.

The big population areas of the world are those
that are calorie short.

To date the substrates for SCP and for fats and
oils by bacterial action have been paraffins, alcohols,
carbohydrates and methane, primarily. One might suggest that synthesis gas, CO plus hydrogen, may be added
in the future.

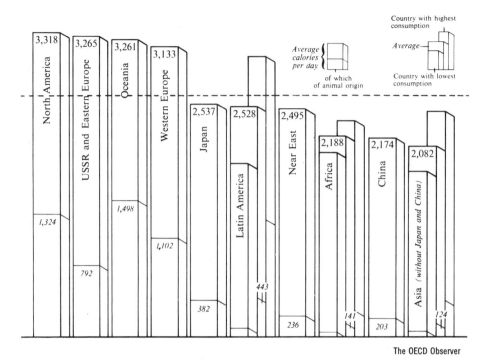

The OECD Observer

Figure 3.   Food consumption per head in 1970 (Source: The OECD Observer, *No. 81, May-June 1976, p 6)*

Two other developments might be mentioned.   One is the bacterial leaching of desired minerals, particularly copper and uranium.   To a degree, naturally occurring bacteria are used, but more and more strains are being developed for specific applications.   They are propagated by fermentation.

An off-shoot of this is the bacterial treatment of agricultural ground.   Russian work has indicated that the application of sulfur, plus bacteria, can make phosphorus available for use by the growing plant in situ, phosphorus present in the soil that otherwise would be unavailable.   This, then could result in a lowering of fertilizer phosphate application.   To our knowledge, nothing has been done along these lines in the U.S.

One of the interesting things about some of these new developments is that they are on a very large scale, that is, compared with many existing fermentations. This has resulted in the development of new types of fermenters.   With the very large throughputs required

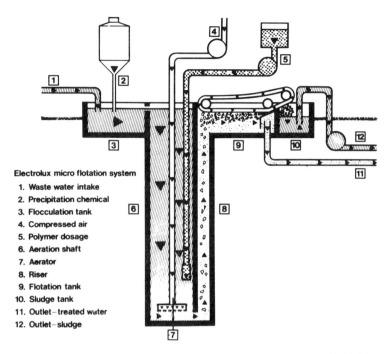

Electrolux micro flotation system

1. Waste water intake
2. Precipitation chemical
3. Flocculation tank
4. Compressed air
5. Polymer dosage
6. Aeration shaft
7. Aerator
8. Riser
9. Flotation tank
10. Sludge tank
11. Outlet–treated water
12. Outlet–sludge

Chemical Age

*Figure 4. Flowsheet for Swedish microfloatation process licensed by ICI
(Source:* Chemical Age, *Jan. 30, 1976)*

for SCP production came the development of air-lift fermenters, in effect adding to the oxygen partial pressure.

From that experience, Imperial Chemical Industries Ltd., has gone a step further and has dug the fermenter down into the ground. It is shown in Figure 4, as now applied to secondary waste disposal treatment. Gulf has come up with an air lift modification called a basin fermenter, as shown in Figure 5. These are engineering modifications for a lower investment and/or new uses.

In the anerobic area there are engineering tricks, such as membranes and operation under vacuum, that are coming into play on large scale. Other tricks-to-the-trade are coming along.

Thus, one now can get the vision of large-scale chemical engineering being applied to fermentations, perhaps for the first time. There will be those who argue that good chemical engineering has been applied

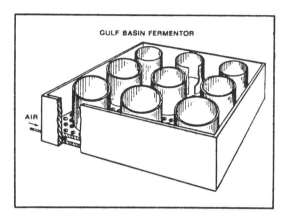

Chemical Engineering Progress

Figure 5. Basin fermentor design (Source: P. G. Cooper and R. S. Silver, Chemical Engineering Progress, *Vol. 71, No. 9, Sept 1975, p 86)*

to fermentation for a long time.   This may be true, but the concept of scale really did not hit until SCP came along and the petroleum companies became interested.   Now the energy people are getting into the act, energy other than just petroleum.   Here we can look at an even larger scale than SCP as a potential, and generally anaerobic rather than aerobic fermentation. What ingenious fermenters will come out of that work still is to be seen.

The point, is that one now looks at the whole unit process of fermentation in a new light.   Whether you are interested in substrates, in fermentation chemistry, or the fermentation industry -- this development is a healthy one.

Abstract

The increase in oil and coal costs, and worry about the long term supply of fossil fuel raw materials have caused a renewed interest in fermentation using renewable substrates.   New developments are occurring both in the biochemistry and equipment facets of fermentation technology.   Organism mutation and adaptation methods make the processing more effective. Continuous and large-scale equipment with improved energy balance widen the potentials for economic production.   Past chemical production using fermentation is reviewed as a prelude to the future.

Biographic Notes

Roger Williams, Jr., Pres.,Chem and Eng. Economics Consultant. Educated at Amherst Coll. and Mass. Inst. of Tech. Industrial experience at the Dupont Co. and Chem. Eng. Magazine; organized his own consulting service in 1950. Roger Williams Technical and Economic Services, Inc., 34 Washington Rd., P.O. Box 426, Princeton, N.J. 05540. U.S.A.

21

Production of Industrially Important Gums with Particular Reference to Xanthan Gum and Microbial Alginate

CHRISTOPHER J. LAWSON

Tate & Lyle Ltd., Group R & D, P.O. Box 68, Reading, Berks, RG6 2BX, England

The production of polysaccharide gums by fermentation has been described by some of the more optimistic microbial technologists as the next major fermentation area. It now is gaining treatment in public and private meetings similar to that offered to single cell protein some years ago. This optimism is based on the undoubted success of the major product in this area, xanthan gum, which has raised the tantalising prospect of a whole range of microbial gums which could not only reflect and improve upon the available plant gums, but also introduce novel properties for exploitation in existing and as yet undeveloped applications. About a dozen major companies are thought to be conducting substantial research and development programmes into the production of microbial polysaccharides; some of them already are in the fermentation industry, but others, like Tate and Lyle and Hercules, are newcomers to this technology. Despite this heavy research and development effort which, in some instances has a history of at least a decade, the state or at least public knowledge of the technology as judged from the patent and scientific literature, is low, to say the least. There is little literature on the production technologies used by industry. Academic microbiology, for the most part, has ignored the physiology of exocellular polysaccharide synthesis and excretion. The physiology of polysaccharide synthesis has been studied as a basis for developing production processes. A number of the results obtained on microbial alginate in the Tate and Lyle laboratories will be discussed. As a brief introduction to microbial gum production, it will be useful to contemplate a number of more general questions which would be posed by an individual or group of individuals considering this field for possible exploitation.

1. What are biopolymers of industrial significance?
2. Why should sugar companies show any interest?
3. What is the commercial status of microbial gums?
4. What are the properties of interest?
5. What advantages has fermentation to offer?
6. What disadvantages does fermentation have?

From the point of view of commercial usefulness, microbial polysaccharides are found as slimy or gelatinous materials secreted into the aqueous environment upon which are grown many bacteria, fungi and yeasts. The reason why this group of compounds has received attention is connected with the rheological and gel-forming properties which they show, defined by comparison with the well-known, industrial gums from plant sources. In Table I the major areas of gum classification are presented, showing comparisons of both plant and microbial gums.

The interest of Tate and Lyle in microbial gums was stimulated by the effective increase in value which might be given to sucrose if used as a fermentation substitute. Generally, gums can be classed as high

Table I.

Classification of Water Soluble Polysaccharides

Origin		Examples
(i) Unmodified Gums		
plants	trees	- gum arabic
	seeds	- locust bean gum
		- guar gum
	seaweeds	- agar
		- alginate
		- carrageenan
	cereals	- cornstarch
	tubers	- potato starch
	citrus fruits	- pectin
micro-organisms	bacteria	- dextran
		- xanthan gum
(ii) Modified Gums		
plants	trees	- carboxymethyl cellulose
	grasses	- methyl cellulose
	cottons	- hydroxymethyl cellulose
	cereals	- dextrins
	tubers	- carboxymethyl starch
	seaweeds	- propylene glycol alginate
	citrus fruits	- low methoxy pectin
micro-organisms	bacteria	- D.E.A.E. dextran (Diethyl amino ethyl)

value chemicals, and it was considered that their production by fermentation processes could lead to returns
on capital employed better than those obtained, for
example, by straight financial investment.   There also
was a natural attraction towards a potential involve-
ment with the new area of diversification, which production of microbial gums would offer.   Glucose and
glucose syrups may be obtained more cheaply than refined cane sugar and, in any logical process development exercise, it would be necessary to acknowledge
this.   For certain polysaccharide fermentations, however, the use of raw sugar, refining syrups or molasses
may be contemplated substantially improving the economics in comparison with glucose.   Also, many sugar
companies have strong intrinsic interests in carbohydrates other than sucrose.   Again, this is a possible
reason for an interest in fermentation substrates other
than sucrose.   One obvious disadvantage for a sugar
company contemplating the manufacture of microbial
gums is the probable lack of both applications know-how
and marketing expertise.   One of the main motivating
reasons for Tate and Lyle's enthusiasm at a joint
venture with a company like Hercules, was in the
strength to be gained from an organisation already
marketing gums.
        A vast number of microorganisms produce exocellu-
lar polysaccharides, and many publications have appeared, in which they are described.   The majority, however, have not been exploited commercially.   In Table
II some polysaccharide-producing microorganisms are
shown to give an idea of the wide occurance of the
polysaccharide producers.   In Table III are listed
those gums which appear to be the most advanced in
terms of their commercial development.   This demon-
strates the commercial status of the gums and gives an
indication of possible future trends.   To give an
idea of the sort of production volumes of microbial
gums, it has been estimated that the production of
xanthan gum by all manufacturers amounted to some
6,000 tonnes in 1975.   Table IV presents for comparison the consumption of gums in the United States in
1973.   Production costs for microbial gums are very
high, mainly due to the high cost of plant.   A number
of reasons for this are given later.   Kelco, for
instance, has announced that their new plant for xanthan gum production in Oklahoma, will cost $35 million.
The prices charged reflect this high production cost
and therefore, it is not surprising that the price of
xanthan ranges from $3.5 per lb to $4 per lb, depending
upon grade.

Table II.

Some Polysaccharide-Producing Microorganisms

Gram Positive Bacteria	Yeasts
Bacillus spp.	Rhodotorula spp.
Leuconostoc spp.	Pichia spp.
Streptococcus mutans	Pachysolen tannophilus
Streptococcus spp.	Lipomices spp.
Gram negative Bacteria	Hansenula capsulata
	H. Holstii
Azotobacter spp.	Cryptococcus spp.
Rhizobium spp.	Torulopsis molischiana
Escherichia coli	T. Pinus
Klebsiella aerogenes	Aureobasidium pullulans
Acetobacter xylinum	Other Fungi
Arthrobacter viscosus	
Pseudomonas aeruginosa	Penicillium spp.
Xanthomonas campestris	Tremella mesentaria
X. phaseoli	
Achromobacter spp.	
Alcaligenes faecalis var. myxogenes	
Agrobacterium spp.	
Erwinia spp.	
Sphaerotilus mutans	

Table V gives an indication of the type of process either being operated or in development. It is thought that all commercialised processes, at the moment, are batch, and all substrates are carbohydrates, mostly based on glucose or sucrose.

In the selection of microbial gums for commercial exploitation,inevitably the conclusion is reached,that, for a gum to be successful, it must have some unique physical property. As already mentioned, production costs are high and, it is unrealistic to contemplate developing gum systems which show non-specific thickening and suspending properties. In Table VI are listed some polysaccharides, both plant and microbial, identifying the physical properties which are unique, and uses which are specific and not easily copied by other gums.

Both the advantages and disadvantages of the fermentation approach to polysaccharide production now will be examined (Table VII). Advantages of fermentation over traditional methods are firstly in medium preparation. The raw materials such as carbohydrate substrates,nitrogen sources and inorganic salts normally are readily available and, in many fermentations, it

Table III.

Microbial Polysaccharides of Commercial Importance   (Commercial Information)

Name	State of Development	Trade Name	Company Involved
Dextran	Present – In production Future – Static	Various	Dextran Products Polydex
Xanthan Gum	Present – In production Future – Expanding	Keltrol Kelzan Rhodigel 23	Kelco Co. Rhone Poulenc/ General Mills
	Present – In development Future – Commercialisation announced	–	Tate & Lyle/ Hercules
Pullulan	Present – In development Future – Commercialisation announced	Pullulan	Hayashibara Corp.
Erwinia Exopolysaccharide	Present – In production (U.S.A.) Future – Not known	Zanflo	Kelco Co.
Scleroglucan	Present – In development Future – Not known	Polytran F.S.	Pillsbury
Microbial alginate	Present – In development Future – Promising	–	Tate & Lyle Ltd./ Hercules
Bakers Yeast Glycan	Present – In development Future – Not known	BYG^R	Anheuser – Busch Inc.
Curdlan	Present – In development Future – Not known	–	Takeda Chemical Ind.

Table IV.

The Consumption of Industrial Gums in The United States (1973)* (Tonnes)

Gum	Food Uses	Industrial	Total
Corn sugars	2,232,142	?	–
Cornstarch	223,214	1,116,071	1,339,285
Carboxymethylcellulose	6,696	43,303	50,000
Methylcellulose	900	23,660	24,553
Guar	6,696	15,625	22,321
Arabic	10,267	3,125	13,392
Pectin	5,357	0	5,357
Locust bean	4,017	1,785	5,803
Alginate	4,017	4,017	8,034
Ghatti	4,464	446	4,910
Carrageenan	4,017	89	4,106
Xanthan	1,000	2,678	3,678
Karaya	446	3,125	3,571
Tragacanth	580	89	669
Agar	133	178	311
Furcellaran	89	0	89

* R.L. Whistler 1974 (unpublished results)

Table V.

Microbial Polysaccharides of Commercial Importance   (Process Information)

Name	Organism	Type of Process	Substrate	Component sugar of polymer
Dextran	Leuconostoc mesenteroides	cell free enzyme	glucose (sucrose)	glucose
Xanthan gum	Xanthomonas campestris	bacterial	glucose glucose syrup	glucose (acetate) glucuronic acid mannose (pyruvate)
Pullulan	Aureobasidium pullulans	fungal	glucose syrup	glucose
Erwinia Exopolysaccharide	Erwinia tahitica	bacterial	glucose? glucose syrup?	glucose galactose fucose uronic acid (acetyl)
Scleroglucan	Sclerotium glucanicum	fungal	glucose	glucose
Microbial alginate	Azotobacter vinelandii	bacterial	sucrose	mannuronic acid guluronic acid (acetate)
Bakers Yeast Glycan	Saccharomyces cerevisiae	yeast	glucose	glucose mannose
Curdlan	Agrobacterium sp Alcaligenes faecalis	bacterial	glucose	glucose

Table VI.

Some Important Physical Properties of Industrial Gums and their Applications

Physical Property	Gum	Application
Cold set, clear, gel formation with divalent cations	Alginate	Re-formed fruit pieces Dental gels
Gel formation with sucrose	Pectin	Jam manufacture
Heat reversible gel formation	Agar	Microbiological solid media Synthetic meat gels
Heat reversible gel formation in the presence of potassium ions	Carrageenan	Synthetic meat gels Instant desserts
Non-reactivity with 'reactive' (Procion) dyestuffs	Alginate	Textile print paste thickener
Stability in the presence of strong acids	Xanthan gum	In rust curing gels containing phosphoric acid
Pseudoplastic behaviour under conditions of high shear	Xanthan gum	As a lubricant for the bentonite muds used to drill oil wells
Synergistic gel formation with carob and guar gums	Xanthan gum	Synthetic meat gels
Retardation of sugar crystallisation at low moisture contents	Gum Arabic	In pastilles and jujubes

is possible to use precisely defined media compositions. In the actual process of fermentation, ongoing parameters such as pH, temperature, fermentation time, dilution rate and aeration can be controlled and manipulated. Two interrelated results potentially are possible through these controls. The first is the maintenance of product specifications within defined limits, and this is particularly vital in polysaccharide fermentations where even small changes in certain process variables can have dramatic effects on polymer structure and, therefore, physical behaviour. This, incidentally, is one reason why the use of continuous culture is favoured. Under steady state conditions, the variables of fermentation are held constant with respect to each other, thus affording a much better understanding and control of the process. The second effect is the potential ability to manipulate product type and yield. For example, the ability through specific changes in fermentation conditions to produce ranges of microbial gums of a particular type, differing in molecular weight. Another advantage of fermentation is that, in product recovery, harsh extraction techniques normally are not required and, very often, solvent precipitation may be contemplated which, of course, is a very mild treatment, not likely to lead to product degradation. The choice

Table VII.

Benefits Obtained in Producing Polysaccharides

by Fermentation

1.   Medium Preparation

(a) Raw materials in plentiful supply

(b) Precisely defined media possible

2.   Fermentation

(a) High degree of process control possible

(b) Continuous culture possible∴ high productivity

3.   Product Recovery

(a) Mild conditions can be used∴little product

degradation.

Table VIII.

Technical Problems in Microbial Polysaccharide Production

High broth viscosities, resulting in :

(i)   Low product concentration, therefore
            large volumes of water
            large fermenter capacity

(ii)   High energy requirements for
            oxygen transfer
            bulk mixing
            water removal

(iii)   Difficulties in cell removal.

of production sites can be fairly flexible, limited
mainly by the availability of water. Finally, the use
of continuous culture can enhance productivities over
those obtained in batch cultures, as cells are grown
always under conditions most conducive to efficient
product formation.
Turning to the disadvantages of fermentation
(Table VIII), they can be summarised as being caused
largely by the highly viscous fermentation broths
encountered. This severely limits the concentrations
of polymer it is practicable to synthesize. The range
30-40,000 cps is not uncommon and no concentrations
of gum above about 4% can be contemplated. It follows,
therefore, that very large fermenters are required in
order to synthesise economically sensible tonnages of
gum. Fermenter sizes of 50-200 m^3 are practical.
Very large volumes of both water and product recovery
solvent are required, so that effluent problems must
be contemplated, unless recycling is to be undertaken.
Other problems lie in the high power requirements
needed to obtain satisfactory broth mixing and aeration
which,if not adequate,can lead quickly to oxygen limitation and lowered polysaccharide productivities. Finally, cell removal, particularly by mechanical means
(centrifuge, filters), is difficult and, although methods based on enzyme digestion of cells (1) and alkali
digestion have been published(2),many problems have yet
to be solved.
    In summary, it is important to point out that, al-
though fermentation as an approach to polysaccharide
production is not without problems, the concept could,
in the future,revolutionize many aspects of industrial
gum production.
    Aspects of the work being undertaken into the pro-
duction of microbial alginate at the Tate and Lyle Laboratories now will be examined. Briefly, alginate is
best known as the polysaccharide obtained from brown
algae such as species of Laminaria and Macrocystis
(Figure 1). It is a linear polymer of β-D-mannuronic
acid and α-L-guluronic acid. The arrangement of the
monomers has been studied by Haug and co-workers (3)
and Rees and co-workers (4) and referred to as the
block structure. That is, homopolymeric blocks of
mannuronic and guluronic acid comprise the so called,
"alternating regions". The flow and gel-forming properties of the polymer in aqueous solution depend on
the proportions of the monosaccharide residues, on
their arrangement and on the polymer molecular weight.
This applies very much to gel formation in the presence
of divalent metal cations as, for example, alginates

Monomers

β-D-Mannuronic acid   α-L-Guluronic acid

Block Structure

-M-M-M-M-M-M-

-G-G-G-G-G-G-

-M-G-M-G-M-G-

Figure 1. Structure of alginic acid

having a high proportion of guluronic acid particularly
as homopolymers, form the strongest and most brittle
gels.   The exocellular polysaccharide produced by
Azotobacter vinelandii has been shown initially by
Gorin and Spencer (5) and by the late Arne Haug and
co-workers (6) to have the same basic structure as
that from algal sources except that a mall number of
hydroxyl groups were acetylated.

       In the Tate and Lyle study of Azotobacter alginate
the objective has been the development of a product
which would compete both in behaviour and economic
terms with the algal materials on sale in world markets.
Early studies in batch culture under the conditions
described by Gorin and Spencer (5), gave poor products,
obtained in very low yield.   Subsequent improvements
were made by growing the organism under phosphate deficient conditions (7) plus other modifications which
increased the consistency index (a measure of flow
behaviour in aqueous solution) from ∿30 cps to 4,000
cps, thus covering the range of commercially available,
algal alginates.   This wide range of product viscosities also has been obtained from continuous cultures
(Figure 2).   The most viscous product has a consistency index of 6,000 cps (1% concentration).   The gel-
forming properties of microbial alginate also were
demonstrated to be similar to their algal counterparts.

In batch cultures, the highest yields of sodium alginate possible to obtain under phosphate deficient conditions, were found to approximate 25% of the sucrose utilised (Figure 3). High respiration in A.vinelandii is a well-known phenomenon and,as a result under certain conditions, much of the sucrose can be utilized in an uncontrolled manner and lost as carbon dioxide. It,therefore,was decided to investigate the effect of respiration on alginate production in continuous culture; a technique having the potential of much

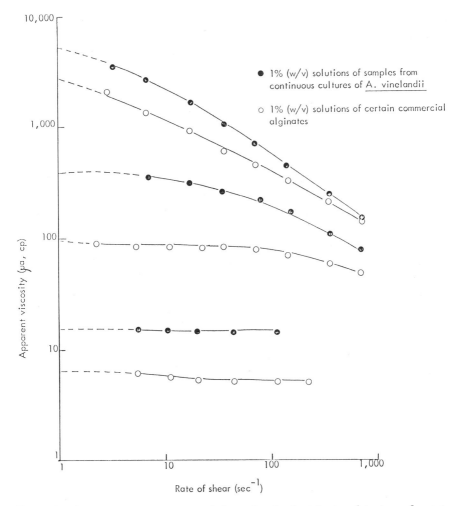

● 1% (w/v) solutions of samples from continuous cultures of A. vinelandii

o 1% (w/v) solutions of certain commercial alginates

Figure 2.   Apparent viscosity vs. rate of shear plots for Azotobacter *alginates and certain commercial algal alginates*

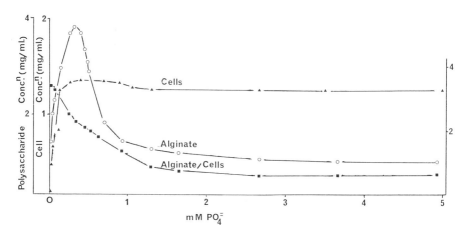

Figure 3. *Effect of phosphate concentration on alginic acid synthesis by* Azotobacter
vinelandii *in batch culture*

greater control. Phosphate-limited growth conditions
were chosen, as a paucity of phosphate was indicated by
the work in batch culture. Respiration rate was controlled by manipulating fermenter impeller speeds,
resulting in altered oxygen transfer into the fermentation broth. Therefore, although cell mass was
controlled at an essentially constant level by the
phosphate availability, the specific respiration rate
was determined by the availability of oxygen (Figure 4).
Under the above conditions, at lower respiration rates,
the maximum yield of sodium alginate was in the region
of 45% of the sucrose utilised. At higher respiration
rates, the yield fell dramatically due to a greater
proportion of sucrose being burned off as CO_2.
     In conclusion, the production of microbial poly-
saccharides of commercial significance is now a well
established fact and a new generation of water soluble
polymers is being identified. However, the production
technologies are expensive, relative to many of the
plant gums. Biopolymers will only gain a fully
established competitive position leading to the use of
industrial scale continuous cultures through the success of current research programmes into the physiology
of exopolysaccharide synthesis.

Abstract

     In recent years, the exocellular polysaccharide
elaborated by the bacterium <u>Xanthomonas compestris</u> has
emerged as a product with significant industrial appli-

cation, and the present annual world consumption is
several thousand tons.   This has demonstrated the potential of fermentation for producing polysaccharides
having unusual solution and gel properties.   At the
present time there is an increasing demand for xanthan
gum and new applications appear regularly in the patent
literature and elsewhere.   It is likely that other
microorganisms will be capable of producing commercially valuable polysaccharides and, as a consequence, a
number of systems are under investigation with several
apparently at an advanced stage of development.   Tate
and Lyle became interested in microbial gums through
reports that the bacteria <u>Azotobacter</u> <u>vinelandii</u> and
<u>Pseudomonas</u> <u>aeruginosa</u> produce polysaccharides similar
to the polyuronide, alginic acid.   The sole source
of this polysaccharide at present is the brown algae
from which approximately 20,000 tons of alginate are
extracted per annum.   <u>Pseudomonas</u> <u>aeruginosa</u> was
rejected for study due to its association with patho-
genic conditions in man.   As initial assessments
indicated that the <u>Azotobacter</u> polysaccharide could be
commercially valuable if sufficiently high yields were

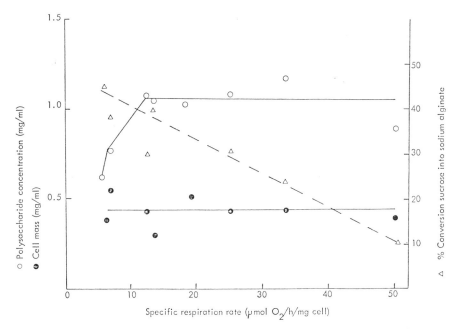

Figure 4.   Exopolysaccharide production by Azotobacter vinelandii *at a range of respira-tion rates*

obtained, the latter was selected for further study.
Subsequent developments to pilot plant level have involved improving yield and controlling and manipulating
the physical properties of the polymer produced by
appropriate choice of growth media and fermentation
conditions, using both batch and continuous culture and
also by strain selection.

Literature Cited

1. Kelco Co., British Patent, 1,443,507, (1975).
2. Patton, J.T., United States Patent, 3,729,460,
    (1970).
3. Larsen, B., Smidsrod, O., Haug, A. and Painter, T.
    Acta. Chem. Scand., (1969), 23, 2375.
4. Rees, D.A., Biochem. J., (1972), 126, 257.
5. Gorin, P.A.J. and Spencer, J.F.T., Canad. J. Chem.
    (1966), 44, 993.
6. Larsen, B. and Haug, A., Carbohyd. Res., (1971),
    17, 287.
7. Imrie, F.K.E., British Patent, 1,331,771, (1973).

Biographic Notes

    Christopher J. Lawson, Ph.D., Project Leader -
Microbiological Polysaccharides. Educated at the
Univ. of Edinburgh. Joined Tate and Lyle, Ltd., in
1969, specializing in microbiological polysaccharides
by fermentation. Tate and Lyle, Ltd., Philip Lyle
Memorial Research Laboratory, P.O. Box 68, Reading,
Berkshire RG6 2BX, England.

Single Cell Protein and the Protein Economy

JEREMY J. WELLS

Biochem Design S.p.A., Via A. Bargoni 78 - 00153 Rome, Italy

The primary aim of this paper is to examine carbohydrates as potential substrates for the production
of single cell protein (SCP) and compare them with
other substrates such as hydrocarbons.

An overview of actual and potential supplies of
protein from various sources is necessary to show the
requirements of SCP in the protein supplement market.
Development of the present status of SCP projects
shows the changes during the past year.

Man judges his standard of living by his consumption of high proteins such as meat, milk and eggs.
To produce these high proteins, requires a large and
complex animal feed industry which today not only is
well developed but highly sophisticated. Protein
supplements are produced mainly because a large market
for them exists. The role of SCP as a protein supplement, has become well established over the last few
years. Large amounts of nutritional and toxicological
data have provided numerous government regulatory authorities sufficient evidence to give product approval
in many developed countries.

It has been established clearly that SCP is an
excellent source of protein for animal feeding. However, at this time, the production of SCP still is not
commercially viable on account of the ample supplies
of proteins available at prices lower than SCP can be
produced.

The hopes for having two major production units in
operation during 1976 have been dampened now by political bans, brought on by Italian consumer groups, on the
two main processors, Italproteine and Liquichimica S.p.
A.

Development of technology to produce SCP from
cellulose remains commercially elusive. However, in
view of the expected life span of hydrocarbon resources,

most experts agree that carbohydrates could hold the
greatest potential for future SCP development, since
they are renewable resources. Yet,the economics of
using carbohydrate substrates often are rather poor on
account of many constraints on their use.

Status of SCP Developments

The promise of a 200,000 ton SCP production during
1976 was shattered by the suspension by Consiglio
Superiore di Sanita in Rome of their previous decree to
Italproteine, authorising the use of SCP in Italy.
This has meant that the plants of both Italproteine and
Liquichimica S.p.A.,although finished, cannot be operated, except for limited, test purposes.
British Petroleum has stated publicly that no
technical reason exists to withhold permission for the
use of their product. Confirmation by weighty factual
evidence submitted from all parts of the world makes
this factual. British Petroleum,with over $60 million
tied up in plant,now is attempting a judicial appeal.
Although the financial prospects are poor for operating
these plants in the current protein supplement markets,
they are expected to yield vital operating data for
future developments. This is important, since several
oil producing countries are studying SCP production as
a means of adding value to their oil resources.
In the last year, several major groups have with-
drawn from SCP developments including Shell Chemical
and the French, CFR group. At the same time, Imperial Chemical Industries Ltd. (ICI) have postponed
authorization for their major production unit. It
even is rumored that they cut its projected size to
50,000 tons before presentation to their Board. On
the successful completion of the ICI plant, depends
potential developments in Iran and Japan. However, the
situation in East Europe is more optimistic. In these
countries, economic viability is enhanced by the anticipated savings in hard currency.
The Government of the USSR currently is discussing
with several processors, the development of a 250,000
ton plant. This makes sense in a country such as
Russian which imports large amounts of feed supplements
for dollars, while n-paraffins are a natural resource.
A joint venture between the Governments of Poland and
East Germany is expected to produce a 50,000 ton plant
using a gas oil substrate. And, the 60,000 ton, joint
Japanese and Romanian plant is almost on stream.
Another oil producer intending to produce SCP, is
Venezuela, where a 100,000 ton plant currently is being

built, based on British Petroleum technology.   It is
to be completed by 1979.   Five other oil producers,
Afganistan, Algeria, Libya, Kuwait and Saudi Arabia are
studying the execution of pre-feasibility studies via
the offices of the United Nations (UNIDO).   It is reported that over $140 million of Arabian OPEC money
may have been put aside for novel, petrochemical developments.

Table I comments on the present status of projects
last reported in 1975 during the American Chemical Society's Symposium on SCP in Philadelphia.

One can generalize, saying that, in a developed
country environment, SCP from hydrocarbons is not commercially viable at this time. However, in certain limited circumstances, in countries with low raw material
costs, with low utilities costs, or where hard currency
is scarce, SCP often can be justified economically.
British Petroleum and ICI have maintained programs
to obtain regulatory permission of their products
through extensive testing.   This has been successful
throughout Europe with the notable exception of Italy.

Potential for Carbohydrate Substrates

There has been a long search for an economic means
of producing nutrients from carbohydrate substrates.
In Germany during World War II, yeasts were produced on
wood sugars obtained by hydrolysis of wood wastes.
However, these uneconomic processes will not be viable
in a free economy.   In the Soviet Union today, over
900,000 tons of fodder yeast are produced from wood
sugars.   Several major processors have been trying
to develop a direct fermentation of cellulose. However,
slow reaction times and low cell yields have caused
most processes to be uneconomical.

Carbohydrates can be classified into two classes
for this discussion.   Saccharides and polysaccharides
will be discussed separately.

There is an obvious attraction in using an annual-
ly renewable resource.   However the economics of pro-
duction often is not appreciated in which every poten-
tial substrate must be considered in terms of:
   - Alternative use
   - Logistics of collection
   - Degree of pretreatment, and
   - Seasonal availability.

Table II lists the possible potential substrates
with their various restraints.

Most carbohydrates, particularly conventional
nutritional ones, have considerable market value.   In

TABLE I

PRESENT STATUS OF SCP DEVELOPMENT

Company	Organism	Substrate	Scale	Status
YEAST				
British Petroleum				
Lavera France	Candida Lipolytica	Gas Oil	20-27,000 ton production unit	Plant being converted to another substrate
Grangemouth	Candida Lipolytica	N-paraffins	4,000 ton unit	Plant operating
Sardinia	Candida Lipolytica	N-paraffins	100,000 tons unit completed 1976	Plant awaiting authorization to start
Liquichimica Biosintesi S.p.A.	Candida Novellus	N-paraffins	100,000 tons unit completed 197	Plant awaiting authorization to start
Amoco Foods Co. (USA)	Candida Utilis	Ethanol	4,000 tons plant completed 1975	Plant operating
Slovnaft Kojetin Czechoslovakia	Candida Utilis	Ethanol	1,000 ton pilot plant 60,000 tons plant under development	Still under development
Roniprot, Romania	Candida Pichia	N-paraffins	60,000 ton product. unit	Completed
Bioproteinas de Venezuela	Candida Lipolytica	N-paraffins	100,000 tons unit to by completed 1979	Being constructed
USSR	Candida	N-paraffins	250,000 tons plant under consideration	Under negotiation
East Germany, Poland	Candida	Gas Oil	50,000 tons plant	Under negotiation

TABLE I (cont.d)

PRESENT STATUS OF SCP DEVELOPMENT

Company	Organism	Substrate	Scale	Status
BACTERIA				
Imperial Chemical Industries, England	Pseudomonas methyl-otropha	Methanol	1,000 tons pilot plant 100,000 tons plant under consideration for 1978	Awaiting Board decision on 50,000 ton plant.
Shell Chemical Co.	Mixed culture	Methane	1,000 ton pilot plant under consideration	Indefinitely post-poned.
Exxon/Nestle	Acinetobacter a nitratum	Ethanol	Pilot plant	Operating for test product.
General Electric (USA)	Thermophitic actinomyces	Cellulose	Pilot plant	Not operating
LSU/Bechtel(USA)	Cellulomonas	Cellulose	Pilot plant to be developed	Awaiting development
US Army Natick Research USA	T-Viride and enzymes produced from Trichoderma Viride	Cellulose	Pilot plant	Operating
FUNGI				
RHM Research England	A3/5 Fusarium	Carbohydrates	Pilot plant	Operating
Tate and Lyle Res. England	Aspergillus niger	Carbohydrates	Pilot plant, several village technology plants under develop-ment	100 ton plant in Belize operating

TABLE I (cont,d)

PRESENT STATUS OF SCP DEVELOPMENT

Company	Organism	Substrate	Scale	Status
ALGAE				
Kohlenstoffbiologische Forschungsstation EV Germany	Scenedesmus acutus	CO_2 and sunlight	Pilot plant	Operating
Sosa Texcoco SA Mexico	Spirulina maxima	CO_2 and sunlight	Pilot plant	Project under construction

any free economy, these raw materials always will be sold for the highest market price available. Hence, the alternative uses and demands for a material often will determine its economic availability. Seasonal availability is another important factor. However, the companies trying to develop this technology have tried to develop simple, fermentation processes which are flexible in design to utilize a range of substrates including wastes from food processing.

The use of simple saccharides or disaccharides to grow fodder or bakers yeast has been common for some time. The carbon atoms are easily assimilable by microorganisms.

Molasses, whey, sulphite liquors and potato wastes all have been utilized successfully over the past few years. However, the alternate uses for molasses for direct animal feeding, citric acid or alcohol production have made it available only at a price. In many cases this price has been too high to justify economic exploitation.

Molasses is mixed with bagasse or bagasse pith for direct ruminent feeding. Over 80% of the molasses in the United States is used in animal feeds. With the rising cost of petrochemical feedstocks, molasses again is being considered as a raw material for acetone and butanol production, and, it should be noted that, of the world food alcohol production of 5 million tons, over 3 million tons comes from fermentation.

Other restraints in use of molasses include the pretreatment required to reduce mineral content and excess suspended matter, low cell yields of about 25% of the weight of molasses used, and seasonal availability of between 3 to 8 months per year.

The growing demand for dairy products coupled with the increasing regulatory restrictions on waste disposal have caused considerable research on whey disposal. However, many other types of processes are available, from drying to fractionation. One joint-venture in the United States is using a bound lactose enzyme. Great attention again is being given to cheese whey as a potential substrate. The yeasts grown on whey give an excellent product of about 60% protein.

Both fungi and yeasts have been grown commercially on sulphite waste liquors. However, the gradual change in pulping from sulphite to Kraft processes has reduced availability to such an extent that most commercial yeast plants in North America have closed down. One commercial plant to produce 10,000 tons of fungal protein has been set up in Finland. Similarly, a process to use starch wastes is operating in Scandinavia.

TABLE II

SUBSTRATE FOR MICROBIAL PROTEIN PRODUCTION

| Material | Availability (1) | | Technical | | Competitive USE |
	Place	Time	Pretreatment	Yield (7)	
Saccharide					
Molasses – cane	Widespread	Seasonal	Simple	0.25–0.3	Animal feeding
– beet	Concentrated	Seasonal	Simple	0.27–0.33	Animal feeding
– corn	Limited	Seasonal	Simple		
– citrus	Limited	Seasonal	Simple	0.18	Animal feeding
Whey	Limited	Year-round	Simple	0.03	Fractionation
Sulfite waste liquor	Concentrated	Year-round	Simple	0.008	(2)
Potato waste	Limited	Seasonal	None		
Fruit vegetable packing wastes	Limited	Seasonal	Nore or simple	0.0.3 (3)	
Polysaccharide					
Starch – grains	Widespread	Seasonal	Hydrolysis	0.5–0.6	Food
– cassava	Concentrated	Year-round	Hydrolysis		
Cellulose – wood	Concentrated	Seasonal	Hydrolysis	0.03(4)	Fuel
– bagasse	Concentrated	Seasonal	Hydrolysis	0.1–0.3	Fuel/Animal feed
– corn cobs	Limited	Seasonal	Hydrolysis	0.13	
– hulls	Limited	Seasonal	Hydrolysis		
– municipal wastes	Concentrated	Year-round	Hydrolysis		

TABLE II (cont.d)

SUBSTRATE FOR MICROBIAL PROTEIN PRODUCTION

Material	Availability (1) Place	Time	Technical Pretreatment	Yield (7)	Competitive Use
Hydrocarbons					
Methane	Concentrated	Year-round	None	0.3–1.4 (5)	Fuel/Chemical feedstock
N-paraffins	Widespread	Year-round	Separation	1.0	Fuel/Chemical feedstock
Alcohols					
Methanol	Widespread	Year-round	None	0.25–0.5 (6)	Fuel
Ethanol	Widespread	Year-round	None	0.6–0.7	Fuel
Propanol (n,i)	Limited	Year-round	None	0.4	
Other					
Acetate, maleate	Limited	Year-round	Dependent on source	0.35	

NOTES:

(1) Terms used under "Availability" are defined as follows:
  "Widespread" – available through the world in potential useful amounts
  "Limited"   – relatively small amounts available at a number of locations within large regions
  "Concentrated"–relatively large amounts available at a few locations.

(2) Sulfite waste liquor availability is decreasing (see text).

(3) Yield figure is for citrus press juice.

(4) Yields for cellulosic materials are based on weights of material as delivered, including normal moisture.

(5) Reported yields on methane vary widely; 1.0 is suggested as a resonable norm.

(6) A yield of 0.4 is suggested as a norm for methanol.

(7) All yields are reported in terms of kg dry cells (biomass)/kg substrate supplied.

(Source: Elmer Gaden – SCP Seminar, Rome 1973)

Polysaccharides (such as cellulose) offer the
largest availability of substrate. Each year photosynthesis replaces up to 22 billion tons. The cost
of cellulose for SCP production must be tied to its
value in paper, animal feed or use as a fuel. Sugges-
tions have been made to cultivate polysaccharides for
substrate usage. However, prices of many of these
nutritional and commercial carbohydrates have risen so
high as to preclude their possible use.

Bagasse utilization is of prime interest to the sugar industry. Microbial degradation has been researched
for many years. However, in this time of ever increasing fuel costs, bagasse remains the prime fuel in most
raw cane sugar factories. In addition, bagasse also
is used to make paper board. New process developments
have lowered the minimum size of plant to produce paper
economically.

New developments from Japan are expected to increase the digestibility of bagasse for ruminent feeding.

The process of the US Army Laboratories appears
very attractive but would take three to five years
to commercialize. The two most advanced processes
appear to be the Louisiana State University/Bechtel
and the General Electric processes. But neither

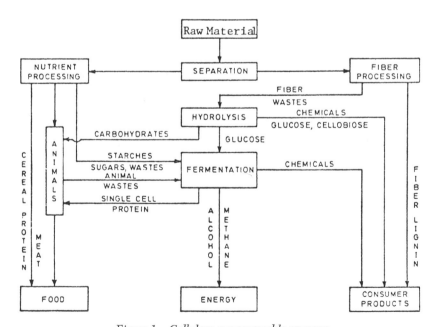

Figure 1. Cellulose as a renewable resource

yet appears ready for commercial exploitation.
A multi-company study in Europe on utilization
of straw as a potential substrate proved to be without
commercial foundation.
The logistics and costs of collecting the straw
prevented the project being justified.
A scheme to utilize cellulose as a renewable resource, developed by Prof. A.E. Humphrey, is shown in
Figure 1. Cellulose from either crops or solid wastes
can be converted into enzymes, proteins, sugar,alcohols
and other biochemical products for food, energy or consumer products. The pretreatment of bagasse requisite
for fermentation also adds to its cost as a potential
substrate. It is concluded that direct fermentation
of cellulose still is several years away.
Considerable effort also is spent on developing
SCP for direct human feeding. At the moment only the
yeast, <u>Candida utilis</u>,produced on sulphite waste liquor
or ethanol, is permitted in foods at inclusion levels
of less than 3%. The only protein isolate produced to
date has been made from bakers yeast and approved as a
food additive by FDA.
Both DuPont, in cooperation with RHM foods in
England, and Nestle/Exxon aim to produce food grade
products by the eighties. Requests for reports of
progress to these companies have yielded replies that
the projects are still in the research phase and pro-
ceeding on schedule.
FDA was asked also to consider refined sugar as
a potential feed stock. On account of the high yields
laboratory work is being carried out on sugar but, due
to the lack of functionality of the protein product,
no market yet is in sight. However, at current price
levels, the cost of digestible carbohydrate automati-
cally would require the microbial product to enter the
high price end of the food protein market.
Current costs of sugar, dextrose and corn syrups
at about 35, 16 and 15 cents per pound, respectively,
indicate that commercial prospects for a feed grade
SCP lie more in the utilization of waste or unrefined
carbohydrate sources.

<u>Potential Markets for Microbial Protein</u>

In order to determine the viability of SCP development, it is necessary to understand how SCP fits into
the protein supplement market and how the market is
projected to develop. At this time, from hydrocarbon
substrates in general, SCP is not commercially viable. With normal paraffins costing in excess of

$200 per ton, SCP would have to be marketed in excess
of $600 per ton to show a return on investment.   These
high prices in a low priced, protein meal market (see
Table III) have not allowed SCP to penetrate the non-
ruminant market.

The original marketing concept had been to develop
SCP with an amino acid profile similar to fish meal to
give it a price premium of 25% on account of the added
growth properties provided by SCP inclusion.   However,
even if a 25% price premium were justified, at today's
fishmeal price of $350 per ton, SCP could not even com-
pete in non-ruminant rations.

A later marketing strategy has been followed to
replace dry milk powder in calf feeding by developing
microorganisms tailored for this purpose.   During
1975, British Petroleum claimed to be selling almost
1500 tons per month into this market.   However, the
French plant of BP now is closed, due to a change of
process from gas oil to n-paraffins, so this market
currently is not being serviced.   Even so, replacement
milk powder in Europe is not politically sound at this
time, due to large overproduction because of the large
subsidies given to milk processors. Table IV shows that

Table III.

AVERAGE INTERNATIONAL MARKET PRICES FOR PROTEIN SUPPLEMENTS

(U.S. Dollars per metric ton)

	1975	1974	1973	1972
Copra	145	160	154	92
Cotton	153	181	219	103
Groundnut	140	187	261	123
Linseed	181	191	227	137
Palm kernel	149	150	149	89
Rapeseed	129	143	178	91
Soybean	155	185	297	130
Sunflower	135	150	217	101
Fish	245	372	542	243

Source: FAO Intergovernment Group on Oilseeds 1976.

Table IV.

PRODUCTION AND STOCKS OF SKIM-MILK POWDER IN WESTERN EUROPE
(1000 metric tons)

Country	Production			End-of-year stocks		
	1973	1974	1975	1973	1974	1975
EEC Countries						
France	730	685	715	203	217	435
West Germany	460	495	530	58	140	346
Netherlands	120	139	184	9	32	146
Belgium	104	115	119	24	51	103
Italy	2	2	2	-	-	-
Ireland	103	109	136	13	35	45
UK	172	106	96	28	29	24
Denmark	52	53	69	1	4	49
Total EEC	1744	1703	1851	334	508	1145
Sweden	49	50	52			
Finland	55	51	50			
Austria	43	42	42			
Switzerland	27	31	34			
Norway	8	9	7			
Total Europe	1928	1885	2037			

Source: IAFMM Executive Council Meeting, April 1976.

carry-over stocks at the end of 1975 were over a million tons.

New legislation by EEC to force the use of milk
powder in animal feeds, instead of alleviating the problem, probably will make it worse.

However, SCP from carbohydrates will aim for a
completely different segment of the market.   Torula
yeast grown on molasses has been marketed for some time
at under $300 per ton.   At today's prices, a feed
grade yeast containing 50% protein would have to be
marketed at under $250 per ton to be competitive.
Tate and Lyle claims for the economics of their Belize
plant are that they will produce a SCP from citrus
wastes for between $160 to $200 per ton, competing with
imported soybean meal at an average price of $270.
This latter price, however, is rather high on account
of the logistics of shipping protein cake to Belize.
Table III gives an idea of the average prices obtained
for the major oil cakes and fishmeal.   Table V shows
the world production of these supplements.   At this

Table V.   Production of Protein Supplements
                   (1000 metric tons)

	1974	1975	1976 (Estimate+
Copra	1235	1540	1620
Cotton	9155	9295	8075
Groundnut	4255	4180	4785
Linseed	1380	1275	1400
Palm kernal	630	655	700
Rapeseed	3870	4165	4425
Soybean	42745	37115	44490
Sunflower	4270	3775	3360
Fish	4150	3910	4125
TOTAL	71690	65910	72980

Source: FAO Intergovernmental Group on Oilseeds, 1976.

time FAO is making projections for 1985 of available
protein sources.
        The biggest unknown to forecast is potential soy-
bean yield.   A 30% improvement in yield over 10 years
would completely alter the structure of the market and
provide an ample supply of protein meal to the year
2000.   It is expected meal prices will increase at a
constant rate of about 6%, in line with inflation.
No large surges are forecast in prices, similar to the
shock condition of 1973.   In general, it is expected
that the rapid growth of the EEC and US feed markets
will not continue at the same pace.   New developing
areas with raw material revenues, such as oil dollars,
are expected to experience rapid growth rates as they
attempt to reach the Western standards of consumption.
The three growth areas for proteins will be the Middle
East, Eastern Europe and South America.
        The vastly increased spending power of the oil
rich nations of the Middle East and North Africa already is causing the growth of protein meal markets.
Within the group of Iran, Lebanon, Iraq, Syria, Turkey
and Saudi Arabia,all are aiming for very large increases in production of meat, particularly poultry, to
achieve better diets for all of their citizens.   Iran
expects to double the 1974 meat production by 1978 to
reach 620,000 tons.   The Iranian Ministry of Agriculture is reported to have under development seven large
feed mills.   In the context of this paper, Iran has a

number of strategically placed sugar mills which could, at this time, be producing microbial protein from waste streams. Such projects could be extremely viable, since a properly placed "village level" plant can be located to provide protein meal in an area unable to obtain conventional feed and still be operated by local labour.

In 1974, Iran imported 55,000 tons of soybean meal but still was unable to meet its Government goals. The market for 200,000 tons of mixed feed in the rest of the Middle East, is expected to reach 1.5 million tons by 1980. Similarly, East Europe will be an excellent growth area. Over $137 million worth of soybean meal was imported to East Europe during 1975.

Food now is considered a strategic weapon, for it is found that an empty stomach leads to discontent. Hence, the USSR and East Europe have considerable programs for more livestock. It can be seen that any SCP product developed from carbohydrates will fall into a price structure defined by its nutritional value, amino acid profile and quality. The market segment is expected to be very large,but the SCP will have to compete for market penetration with conventional products on a price and cost effectiveness basis.

Conclusions

It can be concluded that markets for feed supplements will continue to increase and that SCP in several forms could win a reasonable share of this market. The use of products from the sugar industry such as sugar,molasses and bagasse as substrates are technically feasible but their commercial viability will depend on their values in alternate uses.

Abstract

The demand for protein in human diets has been increasing drastically as a natural result of rising standards of living in all countries of the world. The development of single cell protein for both animal and human food incorporation has been under development for a number of years but has not yet reached the commercially viable stage to justify full scale production. The objective of this paper is to discuss the place of SCP in the range of protein rich feeds and foodstuffs. It also discusses the market perspectives of SCP products and studies the role of sugar and molasses as potential substrates, comparing them to hydrocarbons and other potential feedstuffs. Discussion also is

312   SUCROCHEMISTRY

made of the status of present projects and regulatory
legislation at this time.

Literature Cited

1. European Chemical News, (1976), April 16.
2. European Chemical News, (1976), May 14.
3. Dunlap, C.E., "Economics of Producing Nutrients
    from Cellulose Food Technology", December,
    1975, 62-67.
4. Humphrey, A.E., "Economics and Utilization of
    Enzymatically Hydrolysed Ce-lulose", Finland,
    March, 1975, 12-14.
5. "Intergovernmental Group on Oilseeds, Oils and
    Fats" Statistical Sub Group, FAO Rome, 8-9
    March, 1976.
6. "International Association of Fish Meal Manufactu-
    rers" Executive Council Meeting Report - April,
    1976.
7. "Protein Resources and Technology Status and Re-
    search Needs (1975)",U.S. Government Printing
    Office, Washington.
8. Seeley, R.,"Current Status of Single Cell Protein"
    American Chemical Society, Philadelphia, April
    6-11, 1975.
9. Vlitos, A.J., "Survey of Potential for the Micro-
    bial and Chemical Utilization of Residues of
    Fruit, Food and Agricultural Wastes", OECD,
    May, 1974.
10. Wells, J.J., "Markets for SCP in Developing
    Countries", UNIDO Experts Meeting, Oct. 8-12,
    1973.
11. Wells, J.J., "Analysis of Potential Markets for
    SCP", American Chemical Society, Philadelphia,
    April 9, 1975.

    Jeremy J. Wells, Consultant in proteins by fermen-
tation; Biochem. Design, SpA. via A. Bargoni, 78,
00153, Roma, Italy.

Organic Solvents by Fermentation

FRANK WYNN HAYES

Roger Williams Technical and Economic Services (U.K.) Inc.,
London WC1R 4JH, England

The title sounds simple enough, but it really involves talking about the past, the present and the future.   In addition, the term "solvents" could cover a
multitude of compounds and this review is going to consider only a few; in fact, the focus is concerned mainly with three - ethanol, butanol and acetone.   The
other word in the title is"fermentation"- so that fixes
the parameters.

The discussion of this subject goes back into the
past because we are talking of an industry that was
supposed to have died in the mid-1950's, but never
received a burial.   In 1964, J.J.H. Hastings, in an
article in Chemistry & Industry said, "...ever since
I can remember, the fermentation industry has been dy-
ing... it is the pharmacist who has halted the coffin
of the fermentation industry."   By "the pharmacist",
of course, Hastings meant the antibiotics industry.
These remarks were significant in remembering the background of Hastings, who moved, with Distillers Company,
from fermentation solvents production into antibiotics.

For 20 years prior to those mid-50's, there had
been an established,and growing, industrial fermentation industry, based mainly on the use of blackstrap
molasses as a substrate.   Then, beginning the hydrocarbon age, there came the petrochemicals, derived from
'the ever-so-cheap barrel of crude'.   The byproducts
of the petroleum industry had unlimited promise for
chemical productions.   Olefins and the polymers were
starting their boom.   Everything grew bigger, very
rapidly.   Everyone took it that the 'cheap barrel of
crude' would, like diamonds, be forever.   As for the
fermentation industry - it was a case of "off with old,
and on with the new".   Hence, the statements like that
of Hastings.

The record, however, would be not quite correct if

it was not stated that, just at this time, there was a
sudden upsurge in the popularity and use of molasses
in stock feeding.   This put up the price of blackstrap
and, concurrently, there were significant rises in
freight rates.   All of this tended to make the industries, based on molasses fermentation, very uncertain
about the future.

But, the result was equivalent to an inefficient
job of sterilizing a medium.   You kill everything, or
there is a focus of infection from which growth can
spread.   There was at least one centre of industrial
fermentation activity that refused to lie down, even
though it had been told it was dead.   It did not lie
down for the very good reason that it was making profits.   What better cause can you have for refusing
that coffin?   So, Hastings was not absolutely correct
in his statement.

It must be interesting, and could be instructive,
to take a closer look at that one operation that last-
ed, and ask ourselves "why?".   It was, and is, a
South African enterprise and the reason that I can talk
about it is that I was part and parcel of it in its
most formative and exciting years.

This undertaking, National Chemical Products,Ltd.,
(NCP),at Germiston in the Transvaal, not only was buck-
ing the general surrender to the tidal wave of petrochemicals, it was offending in another, and even more
rebellious manner.   It was not following the worship
at the altar of the god of LARGE SCALE.   Its substitute religion was simply that of FULL UTILISATION.

By this we mean full utilisation of all the properties and components of the feedstock and the side-pro-
ducts that arose in processing.   There could be NO EF-
FLUENT.   The principles involved were:
    Close integration of all the processes in the com-
        plex;
    Greatest possible increase of product value;
    Shared costs for labour and utilities and an
        umbrella of overheads;
    The closest possible energy balance, with full
        utilisation of heat-drop for production of
        power and process steam; and
    Scaling the production to meet requirements of
        known and profitable markets.

Before detailing what was done, and why it was
done, let us restate our case that, not only is there
nothing new under the sun, but that we should be un-
wise to turn our backs on the lessons of practical
experience of the past, even though those days may have
lacked today's technological finesse.

We should divert here, to a particular outlet for
the photosynthetic activities of which we are talking.
Putting the clock back, still in the early 1950's in
South Africa, we were debating seriously the relative
merits of Reppe chemistry and the potential stemming
from a nucleus of fermentation ethanol. Remember that
we had several things going for us - we had the fossil
reserves, in the form of cheap, easily mined coal: we
had the sunshine, the soil, the rainfall to produce
good yields of sugar cane, hence ethanol by fermentation. We were turning our sights towards the new-age
things like ethylene and polymers. We asked advice
from those in the world's chemical centres - who knew.
We got it. We were producing ethanol, by fermentation
at well under £30/ton (then, say, $90/ton). We were
told - "So long as you can do this, ethanol must remain
your best source of ethylene". There were hot debates
about the pros and cons of acetylene vs. ethanol for
many purposes. We had a well-based, efficient carbide
industry, and we therefore, could produce intermediates
like acetaldehyde and crotonaldehyde by either route.
     In Natal, the site and centre of the sugar indus-
try, there was another activity which we must describe
in some detail because, not only is it relevant to the
subject of this talk but, it is very apposite to the
burning questions of today regarding energy sources and
liquid fuels, particularly fuels for the internal combustion engine. In, or around, 1936, C.G. Smith & Co.,
a leading sugar producer with its own estates growing
cane south of Durban, went into fermentation alcohol
production and marketed a motor spirit that was called
"Natalite". This contained some 60% ethanol (Say,
96.5% strength) and 40% benzene. There were problems.
The water in the industrial alcohol used in the mixture
could come out of solution and cause problems in car-
buretion and the engine fuel lines. But, in spite of
this, it worked, it sold, and was quite extensively
used in farm tractors, as well as heavy duty trucks and
ordinary automobiles. The Fordsons and the Fords of
those days seemed to thrive on it. Natalite sold
at a premium of 3-4 pence a gallon compared with imported petrol (gasoline), mainly because of saved duty.
However, even on the small scale of its production, by
the fermentation of blackstrap, plus benzene from the
Natal coal carbonization plants, it was competititive.
Then, after World War II, technology advanced. Usines
de Melle,of France, introduced their azeotropic distillation process, using a benzene entrainer, and so absolute alcohol was an industrial practicality. In 1942,
Natalite was replaced by "Union Spirit", which sells

in and around Durban, Natal, to this day.   "Union"
is very interesting and, in our opinion, should be
examined closely on a world-wide basis.   It is a
straight blend of between 50% to 59% absolute alcohol
with a low-octane gasoline.   In getting it on the
market, the C.G. Smith Group adopted a very progressive
sales policy.   They educated the motoring public to
the fact that this was an alcohol fuel and it had advantages, but was different.   It could affect the
rubber of the fuel pump diaphragm and so, free of
charge, a Union Spirit mechanic would change the diaphragm for one fabricated from a synthetic material,
resistant to alcohol.   Because of the atom of oxygen
in the alcohol molecule, the "Union" required a lower
air/fuel ratio to give a correct carburetor mixture.
Therefore, the carburetor would be adjusted by fitting
a larger jet or a smaller choke tube.   In this way
excellent fuel consumption figures were obtained.
        Union Spirit was not the only alcohol-containing
motor fuel marketed in South Africa.   In 1934, the
Satmar Company started mining torbanite oil shale in
the Transvaal and refining the oil obtained from the
retorting.   At first this was simply mixed with gasoline and sold in the Transvaal as Satmar petrol, again
at a price premium.   In the mid-40's the composition
was changed to approximately 50% imported cracked
crudes, 25% spirits from torbanite oil, 25% absolute
alcohol.   The supplier of the alcohol, made by fermentation, was N.C.P., also in the Transvaal, whose
operations will be described in detail later, but were
centred around the fermentation of blackstrap molasses.
Satmar enjoyed expanding sales until the opening of
Sasol in 1955.   Sasol, in the Orange Free State, whose
activities would require a separate paper to give even
a broad outline, entered the market with its hydrocarbon fuels obtained by the pressure gasification of
coal, with a Fischer-Tropsch synthesis partly by Arge
and partly by Kellogg-fluidized bed.   Sasol Marketing
Co., was formed.   It amalgamated the interests of the
other suppliers and set up its pumps, mainly in the
Transvaal, to supply motor spirit containing the synthesis products from coal, plus the distillates from
imported crudes and naphthas.   The Sasol motor fuel,
sold in the northern Free State and south-western
Transvaal by Sasol Marketing Co. represents only 8%
of the total motor fuel production of Sasol, the balance going to the oil companies for admixture with
their products and distribution through their own
pumps.
        The details of the Sasol activities have been giv-

en only to illustrate the point that, under some circumstances, there is ample scope and opportunity for a
product/market, such as ethanol/motor spirit where the
ethanol is made by fermentation, to have an easy co-existence with motor fuels from other indigenous sources.
     In the case of Union Spirit, still sold today,
still containing over 50% fermentation ethanol, it has
been carried on by an ordinary trading company, in
competition with the other available motor spirits,
without any other motive than that of making profits
for the shareholders.
     So often, when people talk of substitute fuels,
the only concept seems to be one of COMPLETE replacement; yet, if one were considering crises, surely it
is a case of "every little helps" and even 10% has
some significance.
     So now let us return to the example of a diversi-
fied, integrated fermentation complex, with a main-
stream production of solvents, that has proved a
commercial proposition in spite of the coming of the
"camel train, carrying its barrels of black gold!"
     At South Africa's National Chemical Products,
the production flowsheet showed the following main
pathways:

Raw Materials	Intermediates	Products
Blackstrap molasses	Ethanol	Absolute alcohol
Coal	Butanol	Industrial spirit
Water	Acetone	Potable spirit
Cornsteep liquor	Carbon dioxide	Acetone
Ammonia	Hydrogen	Butanol
Lime	Acetaldehyde	Fusel oils
		(amyl alcohol)
Sulphuric acid	Acetic acid	Spirit vinegar
		(10%)
Miscellaneous	Crotonaldehyde	Liquid CO_2
Chemicals		
		Dry ice
		Acetic acid
		Ethyl & Butyl
		acetates
		Cellulose paint
		thinners
		Animal feed
		supplements
		Froth flotation
		reagents

The vital point of this illustration,which is presented

only as being typical, is the co-existence and complete interdependence of all these productions.   There
were two main fermentations   (see Figure 1) - an ethanol fermentation using the usual S. cerevisiae; NOT run
continuously but with the large batch fermenters being
"set" in three separate stages so as to even out the
CO_2 evolution for the benefit of the Dry Ice Plant. In
practice, it proved a good system.   The other fermentation, using Cl. acetobutylicum, produced butanol,
acetone and ethanol in the proportions of 6:3:1, with
CO_2 and H_2 as the gaseous products.   The mixed gases
went on to an absorption plant where the CO_2 was
removed, to be added to that from the ethanol fermentation for the Dry Ice plant and the separated H_2 was
used in an associated phthalic anhydride plant.
     In the earlier operations, the spent wash from
both ethanol and butanol fermentations was evaporated
to between 40 and 45% solids and then incinerated in
a purpose-developed, spray dryer-cum-incinerator in
which the organic matter supported its own combustion,
on much the same principle as a black liquor furnace,
yielding an ash with about 35% K_2O content, which was
sold to the fertilizer compounders.   Then it was found
that a better return could be obtained by separating
the two different types of spent wash from the ethanol
and the butanol fermentations.   The butanol fermentation residues contained vitamins of the B group, so
this material, after concentration in a multiple effect
evaporator, was dried on double-drum dryers to yield a
dry powder with the following interesting composition:

Moisture	3.5 %	(by wt)
Loss on ignition	63.2	
Protein	11.0	
Riboflavin	45	µ/g
Pantothenic acid	175	"
Choline	3200	"
Nicotinic acid	89	"
Pyridoxine	51	"
B 12	0.156	"

This was immediately saleable to feeding-stuffs compounders at a price that started to make the material a
product in its own right,not just an effluent disposal
credit.   In the meantime, the ethanol-spent-wash-ash
continued to sell as a potash fertilizer.   The next
development was the discovery, after much biological
testing and long-term feeding trials, that the combined, spray dried residues from both fermentations, if
incorporated in the rations of grazing animals, had a

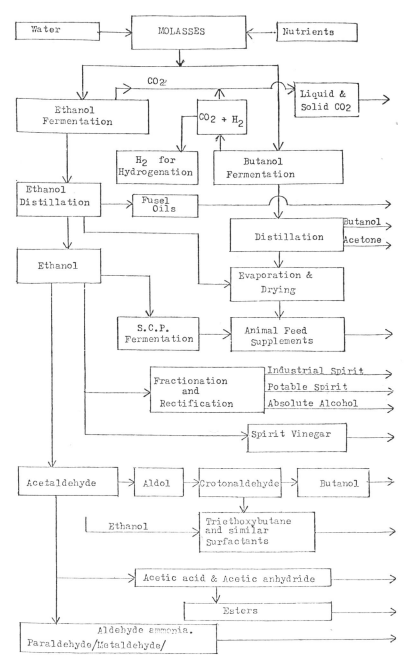

Figure 1

marked beneficial effect on the microflora of the rumen,
enabling the animal to utilize successfully, quite inferior types of straw,grass and other cellulosic materials.   The name of "Rumavite" was given to this
material and it sold on a world-wide basis.
     This, again, is the story of the successful pro-
duction of SOLVENTS by fermentation.
     Some of the ethanol output was converted to
acetaldehyde for captive use.   Acetic acid was made
for the production of esters and also for conversion to
crotonaldehyde which again was condensed with ethanol
to make triethoxybutane, one of a whole range of alkoxy compounds that were patented for use as froth
flotation reagents, particularly successful in copper
and uranium recovery.   And so the story continues,
with each product a logical part of the sequence, with
a carefully investigated market, with a helpful spread
to the fixed charges and general overheads.   In other
words, each being a successful part of the whole but
undoubtedly, if separated and expected to stand as a
production on its own, too small to survive.
     Now what of the future?   Is there a prospect for
the re-institution of fermentation in the production of
solvents?
     The answers to these very "leading" questions are:
1.   Study and use the experience of the industrial
fermentations that worked before the petrochemical age;
more particularly, those which remained alive through
the fact that they made profits.
2.   On the foundation of 1. apply all the benefits
possible by reason of new technologies and techniques
now available to us; with the collaboration of the
disciplines of the geneticist, the molecular biologist,
the biochemist, the microbiologist and the biochemical
engineer.   Now, we can manipulate the microorganisms
of the specific culture that are our working catalysts,
their metabolic pathways, and the enzymes they produce.
By biogenetic manipulations they can be made to do our
bidding; we do not simply have to take the behaviour
pattern of an organism under one set of circumstances
as being fixed and inviolable:we can alter the organism
to suit the desired product, the available substrate,
the fermentation conditions.
3.   If we think in terms of "Large-Scale" for the production of solvents by fermentation then the most important factor is the form, source and cost of the
substrate carbon.   In our opinion there can be only
one sound basis for this renewable raw material supply:
it must be purpose-grown.
4.   If we think in terms of a fairly modest start and

a steady and planned scale-up, then we look very closely indeed at the experience of the past, where diversification, integration and full utilisation of every
constituent of the feedstock and of every intermediate,
are explored.
5.   We cannot talk of a return to "renewable sources"
and fermentation as the chosen route unless we keep always in mind the Energy Balance.   This is the key.
In the growth of crops for a specific purpose we will
be using solar energy, photosynthesis, the Green Leaf -
COMMERCIALLY.   If we should, for instance, grow more
sugar cane, we must use every gram of every stick to
the maximum effect.   If we burn all, or some of, the
residual fibre and cellulosic portion to provide the
heat and power for the process, it must be because we
have no other fuel that is available and cheaper, to
burn.   If there is a suitable substitute fuel, then
we use our fibre from the cane as a source of pentoses
and hexoses.   We use the sugars available to us
according to the product market indications.
6.   We will have to generate steam at high pressure
and utilise the heat drop through a back-pressure or
pass-out turbine to generate power and to supply the
low-pressure process steam.   No loss of latent heat
can be afforded or tolerated.

All the above points, and so many more, belong to
that fascinating subject "FERMENTATION".   There are
technological projections of the new look in fermentation embodying all the factors we have been discussing
and more besides.   A model design can produce solvents
on a large scale.   It involves continuous fermentation,
employing thermophilic organisms.   Distillation requirements are reduced by a very substantial margin.
All components of the substrate are utilized except the
water.

Abstract

Do the escalated prices for a barrel of crude oil,
and the visible end of fossil supplies of energy,justi-
fy a new look at fermentation for bulk chemicals?
They do, but the examination must be thorough, factual
and not simply wordy.   The feedstocks are important,
but just as critical is the energy balance.   So too
are the new tools available to the fermentologist in
biogenetics, biochemistry, and biochemical engineering.
Fermentation technology has not stood still.   But,
renewable sources do not necessarily mean economically
viable sources.   Market surveys plus past experience
can help us in our assessments.   In the latter cate-

gory, experience in both South Africa and Australia
is valuable. The former is somewhat unique because of
its pioneering both in the large-scale use of high concentrations of ethanol in motor fuel and the later development of "oil from coal" by Fischer-Tropsch, with
products and side products that competed with its local
fermentation production of ethanol, butanol, and acetone.

Biographic Notes

Frank W. Hayes, FRIC, FI Chem. Eng., M.S.A. Chem.
Ing.,V.P. Chem. and Eng. Economic Consultant. Educated
at Natal Technical Coll.and Natal Univ.Coll.   Research
chemist at South African Sugar Assoc., Expt. Station
and Chief Chem. Huletts S.A. Refiners Ltd., then
Tech.Dir. of National Chemical Products Ltd., of South
Africa, specializing in organic chemicals by fermenta-
tion; moved to the United Kingdom and the Distiller's
Co., Ltd., Humphreys and Glasgow, Ltd., thence to
Roger Williams Technical and Economic Services (U.K.)
Inc. 37 Sussex Square, Hyde Park, London, W.2. England.

Discussion

Professor Hough: Why is pentaerythritol not utilized? Is it too expensive?

Professor Frisch: Indeed, it is one of the important materials for rigid urethane foams; being used as the polyol. It is used mostly for rigid foam applications. Some of it is also being used in the urethane coating area in the form of polyols. The costs of these polyols are about 45 to 46¢ a pound. They command more or less a premium price as compared to sucrose or sorbitol-based polyols.

Dr. Fuzesi: The cyclic structure - which is missing in pentaerythritol and which is present in sucrose - is very important in the rigid urethane foam area. There the pentaerythritol-based polyol alone, just does not have good foam properties and should be blended with something else. It should be blended with sucrose-based polyol.

Question: Does sucrose have an advantage over oligo-saccharides which have a high molecular weight?

Dr. Fuzesi: It is more difficult to handle the oligosaccharides. Another factor is the availability of sucrose. Mono-, or disaccharides - like the dextrose and sucrose based systems, are well developed and I think those are most suitable for use in rigid foams.

Professor Hough: Therefore, sucrose has many advantages.

Dr. Fuzesi: Yes, many, many advantages.

Question: Do you see any noticeable difference

between the products from sorbitol or lactitol-type
polyols versus sucrose which is not attributable to the
bicyclic character versus the linear polyol?

Professor Frisch: I think it was mentioned that
the bicyclic structure gives it better dimensional stability and better humidity resistance, which is what
you need in a rigid foam.   Those who buy rigid foams,
probably are buying insulation value.   If you do not
have those two properties, you will not be able to keep
your insulation value.   The structure is very important for humidity aging.   The only place I can think
of where you put a rigid urethane foam that you do
not have humidity is in the freezer section of the re-
frigerator cabinet, in between the inner liner and the
outer liner.

Business Aspects

Introduction

G. N. BOLLENBACK

The Sugar Assoc. Inc., 1511 K St., N.W., Suite 1017, Wash., D.C. 20005

In this, the final Session of the Symposium,there will be discussed another very important aspect of the application of sucrose, namely the promotion of sucro-chemistry to the chemical industry.

The presentations of this Symposium have included much discussion on the basic chemistry of sucrose. There have been reviews of the potentials of sucrose-based surfactants where, for the present, it seems the market is to be restricted to special food and pharmaceutical applications because the processes for preparing them seem too expensive to allow the products to be competitive for the general detergent application.

In the field of surface coatings, sucrochemicals encounter the problems of an industry trying to adapt to technological changes, and the long standing, frequently insurmountable, hurdle of convincing commercial formulators to thoroughly evaluate sucrose-based products. To the present there has been but very small penetration into the coatings market using sucrose products as minor adjuncts.

Further, studies of the promotion of sucrose as a polyol, basic to the highly competitive, difficult to penetrate urethane foam industry, disclose some applications, particularly in the rigid foam part of the business. There were presented reasons to consider that it is time for another look at sucrose as a fermentation base and to ponder the great potentials for sucrose in this applied area.

Some pretty hard questions have been put in this Symposium, many centering on the economics and the potential of sucrochemicals. In this final session, perhaps some headway can be made in answering, or trying to answer, some of those questions, as we present these four discussions of the business aspects of sucrochemistry, its processes and its products.

24

Licensing Programs for Sucrochemical Inventions

WILLARD MARCY

Research Corp., 405 Lexington Ave., New York, N.Y. 10017

Sucrose is a valuable component of everyone's diet and is utilized by the body to produce energy for every day living. Its many derivatives, however, have found little use either for foods or as chemical building blocks. Since the sucrose molecule has a number of reactive primary and secondary alcohol centers and can be transformed easily into forms or moieties having anhydro, aldehydic, ketonic and acidic properties, it should be an attractive chemical raw material from which a variety of useful products could be obtained. In addition, sucrose is widely available in very large quantities, being obtained from ecologically attractive natural sources, sugar cane and sugar beets. Its price, in pure form, however, is relatively high and subject to rather severe fluctations due to environmental, economic and political influences.

Because sucrose contains these many reactive centers, nearly every chemical reaction to which it can be subjected results in product mixtures containing large numbers of lesser desired compounds. This situation, in turn, makes it difficult and costly to separate products with reproducible properties from the reaction mix. In addition, a large percentage of the original sucrose molecule inevitably is wasted. Some derivatives of sucrose are edible and metabolizable. These derivatives can perform a number of useful functions relating to other foods and medicine. In these uses sucrose derivatives must be shown to be efficacious and non-toxic to the satisfaction of the Food and Drug Administration authorities. This requires the performance of large scale animal and clinical testing at high costs and much time. Nonetheless, the possible use of sucrose as an attractive commercial raw material had intrigued a number of scientists for a long time. One of these, Dr. Henry B. Hass finally had an oppor-

tunity to exploit his concepts when in 1952, he was
elected President of the Sugar Research Foundation,
now known as the International Sugar Research Foundation, Inc.   That year dates the birth of "sucrochemistry", a word coined by Dr. Hass to describe this new
investigative discipline.

Basic chemical research, best done in academic
institutions or by contract research organizations, was
needed first in order to obtain sucrochemicals of the
nature envisioned by Dr. Hass.   Thus, early work in
sucrochemistry was funded by the Sugar Research Foundation in such research laboratories.   An exploratory
project oriented towards discovering the fundamental
chemistry of sucrose and other carbohydrates had been
supported previously at the Massachusetts Institute of
Technology from 1943 to 1950.   The main purpose of the
project, however, was to train carbohydrate chemists
and to apply classical carbohydrate chemistry more
broadly to sucrose and its close relatives.   None of
the derivatives produced and studied were found to be
of practical importance in later sucrochemical studies,
although a number of patent applications were filed and
patents issued.   Some of these studies were continued
in several other laboratories during the early 1950's,
and a number of additional studies were undertaken to
elucidate the relative reactivities of the various reactive entities, determine reaction kinetics and to
try to make new and possibly useful derivatives.   An
intensive search was undertaken to find mutual solvents
for both hydrophilic sucrose and hydrophobic reactants.

Sucrose Esters and Licensing Policy

Sucrose esters were the subject of the first major
sucrochemical studies starting in 1952.   It was felt
that these materials might be good detergents, easy to
produce inexpensively, non-toxic and biodegradable.
Consequently, a large expenditure was made in several
laboratories and a number of commercially attractive
compounds were produced.   Beginning in early 1955,
patent applications were filed in the United States
and over 20 foreign countries.   These contained claims
both to products and processes.   The patents that
issued were assigned to the Foundation and became the
basis for the first intensive licensing effort.

The mechanics of licensing sucrochemicals basically is no different from licensing other chemicals.
However, special problems in licensing these materials
arise from the nature of their chemical structure,
their method of manufacture, the commercial uses to

which they can be put and the character of the potential licensees.

As the sugar industry traditionally is geared to
produce sugar from agricultural raw materials for
direct edible consumption, little or no attention had
been paid by the individual companies in the industry
to the chemical conversion of sugar to sucrochemicals.
Similarly, since the chemical industry had not contemplated using, nor did it have direct access to sucrose
as a raw material, chemical companies, likewise, never
have developed sucrochemical products in any appreciable way. Thus, when processes for commercially producing sucrochemicals became known, neither the sugar
industry nor the chemical industry was in a position
to utilize the new knowledge. Both industries required a massive educational effort, one which still is going on.

In addition, while Sugar Research Foundation had
given considerable thought to patent ownership and a
patent licensing policy, firm, well thought out imple-
menting procedures had not been established. Some
practical experience, however, had been obtained from
the patenting and licensing of a sterile invert invention discovered under Sugar Research Foundation funding
in the late 1940's. This material was suitable for
intravenous feeding, and patents had been applied for.
Keen interest in obtaining licenses was shown by several drug companies and a license was issued to Baxter
Laboratories in 1950,

This license gave the Foundation an opportunity to
use a patent and licensing policy first developed in
1947. This policy stated that the Foundation should
own patents resulting from research work sponsored by
it, should charge royalties for any license under any
patent it owned, and should license any responsible
applicant on a non-exclusive basis. The policy was
thought to serve two purposes: 1) to ensure that the
results would be used to further the interests of the
industry and to benefit the public, and 2) to obtain
full exploitation of the discovery and give ample
opportunity to demonstrate its usefulness in practice.

The license to Baxter, however, was made exclusive
for 4.5 years and nonexclusive thereafter, until the
expiration of any issued patent. The rationale for
the limited exclusive period was that such protection
was necessary and desirable to induce Baxter to invest
the substantial time and money necessary to develop the
product and create a market. Baxter, subsequently requested an extension of the exclusive period which was
eventually granted for expiration in December, 1957.

The license then became nonexclusive and remained in
effect until the patent finally expired in 1973.
Baxter's sterile invert product was quickly accepted
and enjoyed increasing sales volume during the life of
the license and the product is currently well established for intravenous feeding.

The practical experience obtained from the patenting and licensing of sterile invert provided a firm
basis for the expected, more extensive patenting and
licensing program in the sucrochemical area. In 1957,
the Foundation revised and refined its policies as a
result of a two-year study by a special committee.
While reaffirming the major tenets of the earlier policy, it was decided to adopt standard license forms for
both domestic and foreign licenses, to prohibit the
granting of exclusive licenses and preferential treatment in nonexclusive licenses to Foundation members,
and to provide for sharing royalties with inventors in
educational institutions but not with Foundation staff
members or employees of commercial laboratories which
may have had funded support from the Foundation. This
revised policy first was employed in licensing the sugar esters and has governed all subsequent patenting
and licensing by the Foundation. It still is in effect
today.

The principal terms of the standard license relating to sugar ester detergents include nonexclusivity,
an initial nominal payment on execution of the license,
a running royalty of 2% of net sales value of products
produced under the license, and a minimum annual royalty payment. During 1956, the first year of licensing
activity, four domestic and three foreign licenses were
concluded even while most patent applications were
still pending. In subsequent years a number of additional licensees were obtained, both in the United
States and in foreign countries, but a number of the
original licensees terminated their agreements.

At the present time, 20 years after the original
filing and the year in which the patents expire, sucrose esters are being produced commercially in Japan
and France under the original sugar ester patents.
Recently, a sugar company in England has introduced
sugar esters made by a different proprietary process,
but licenses from Sugar Research Foundation are not
required.

The major licensee in the United States, after
spending a great deal of time and effort on technical
research and development and marketing research, was
unable to obtain Food and Drug Administration clearance
for its contemplated products and discontinued its in-

terest in producing marketable products.
    While it still is possible that widespread use of
the sucrose esters will develop in the United States
in future years, the two major deterrents which will
need to be overcome are costs relative to competing,
petrochemically-based products and FDA clearance for
certain food and non-food uses. The latter problem
arises as a result of the solvent used in the manufacture of the esters. In Japan, both these factors are
less important and, in France, the use is in animal
feed additives where only small amounts are required
and clearance is not necessary.
    It also has been discovered that, unlike the com-
peting fatty alcohol sulphonates, the sugar esters hy-
drolyze extensively during the crutching operation used
in formulating detergents. Further treatment of the
esters with propylene oxide is claimed to alleviate
this problem, but additional research and development
work is needed to prove this.

Sucrose Based Polymers and Other Derivatives

    Since sucrose might seem to be an ideal monomer in
polymerization reactions, a very large variety of reactions of this type have been investigated in Sugar
Research Foundation supported projects. In the 1940's,
ethylene oxide was combined with sucrose, but the products obtained did not lend themselves directly to
commercial development. The first intensive efforts
to produce polymers occurred concurrently with the sugar ester detergent activities in the 1950's. These
efforts included studies of the polymerization of sucrose with urea, vinyl acetate,phenol and formaldehyde,
ammonia and hydrogen, melamine and formaldehyde and
many other variations.
    Polymeric materials based on sucrose also have
been produced by utilizing the free reactive groups of
sucrose partial esters, acetals and ethers. Copolymerization of these derivatives with other esters, acyl
or anhydro derivatives, diepoxides, various vegetable
oils, and cyanoethyl ethers have produced drying oils,
textile finishing agents, adhesives, and polymers suitable for forming into films and fibers. Sucrochemicals, such as allylsucrose and sucrose carbonate, can
be made to form hard surface coatings. Sucrose-based
polyurethanes have received also extensive investigation.
    A number of metal derivatives of sucrose have been
made and tested as antimicrobials,pesticides and fungi-
cides. Some of these also have been tested as chelat-

ing agents and as intermediates in the production of
other ethers and esters. Polyhalogenated sucroses
have been made and their uses per se and as chemical
intermediates have been investigated. Both oxidation
and reduction of the sucrose molecule have been studied,
as well as other derivatives of sucrose, the sulfates,
phosphates and sulfonates.

Many sucrochemicals are akin to fine organic chemicals, pharmaceuticals and organic intermediates. The
major economic attractiveness of sucrochemicals both to
producer and user, however, would be for large volume
chemical products, such as detergents, animal or human
foods, plastics, films or fibers. For these uses,
sucrochemicals would compete with similar and analogous
intermediates and chemicals made from petroleum, a
situation which puts sucrochemicals generally in an
economically unfavorable light. For successful competition with petrochemicals, sucrochemicals must have
special properties or be useful in special ways. This
limits their volume of sales and attractiveness as profit generating products.

Patenting and Licensing Sucrochemical Inventions through Research Corporation

As the number and type of sucrochemicals resulting
from Sugar Research Foundation Studies proliferated and
became more complex, the cost and staff time spent on
patent and licensing matters also increased substantially, and the need for additional help became apparent.
Consequently, in 1960, Sugar Research Foundation and
Research Corporation, a non-profit, tax exempt foundation with considerable prior experience in these areas,
concluded a patent assistance agreement. This arrangement provided that evaluation, patenting and licensing
of inventions arising as a result of Sugar Research
Foundation support were to be undertaken by Research
Corporation at its expense. Any income resulting from
this activity would be distributed between the two
foundations and the inventors. The existing licenses
covering sterile invert and sugar esters were transferred to Research Corporation and future efforts to license these and any other sucrochemical inventions were
to be handled by Research Corporation.

Early in the 1960's, however, due to a policy
change at Sugar Research Foundation, it was decided
virtually to discontinue further general support of
sucrochemical studies, at least temporarily. Nonetheless, studies of a number of polymerization reactions
were continued in Great Britain and Canada, supported

principally by funds provided by industrial members
in those countries. The products and processes resulting from their work, on evaluation, did not appear to
lead to patentable or commercially attractive products.
The level of experimental work was increased somewhat in the first half of the 1970's and patent applications have been filed on four of the most recent developments. Of these, one probably is economically
impractical, one is marginally attractive commercially
and the other two eventually may lead to marketable
products. Licensing of these probably will follow
the general pattern provided by the sugar esters with,
perhaps, the added feature of a limited exclusivity
to encourage investment of adequate funds and technical
support by the licensees.

Reviewing the last twenty years of work, it seems
apparent that, while the chemistry of sucrose is intriguing scientifically, the fate of sucrochemicals in
the marketplace depends on economic and political
factors and competition from petrochemically-based products. If and when petroleum becomes scarce and costly as a raw material, sucrochemicals derived from a
natural, ever-available agricultural source probably
will be able to fill market needs quite well. In the
meantime, these materials will find limited uses in
specialty markets, probably at slowly increasing rates,
thus providing licensing opportunities for the industry
supported, International Sugar Research Foundation,Inc.
and licensing, manufacturing and marketing opportunities for the sugar industry itself.

Abstract

Sucrochemical patents present special licensing problems for several reasons; (1) chemicals based on sucrose usually are complex mixtures rather than pure
chemical compounds, (2) they frequently can be used in
foods or for cosmetic formulation thus requiring FDA
clearances, (3) the same materials can be used for a
wide range of applications such as animal feeds, human
nutrition, polymers, coatings, agricultural chemicals,
wood impregnation, antifungal and antibacterial agents
and (4) they are costly compared to similar competitive
petrochemical derivatives. This paper discusses both
successful and unsuccessful licensing efforts by Research Corporation and the International Sugar Research
Foundation, Inc. over the past 20 years, involving sucrochemicals developed under the long-term research
program supported by that Foundation.

Biographic Notes

Willard Marcy, Ph.D., Vice Pres. - Patents. Educated at Mass. Inst. Tech. Became Asst. Supt. at the Amstar Corp. in 1937, rejoined Amstar as Chem. Eng. R & D in 1949, left in 1964 as Head - Process and Dev. to become Dir. and Vice Pres. - Patents. The Research Corp., 405 Lexington Ave., New York, N.Y. 10017 U.S.A.

25

Can Sucrose Compete with Hydrocarbons as a Chemical Feedstock?

W. J. SHEPPARD and E. S. LIPINSKY

Battelle Columbus Laboratories, 505 King Ave., Columbus, Ohio 43201

Many chemical uses for sucrose and its derivatives have been reported including the making of ethyl alcohol, glycerin, urethane foam intermediates, detergents, and plasticizers. These opportunities have been discussed in other papers in this volume, in the ISRF-published book by Valerie Kollonitsch(1), and elsewhere.

With the rise of oil and gas prices that have occurred recently and that are expected in the future, updated comparisons are needed of costs of making various chemicals from hydrocarbons versus sucrose. In the long run, oil and gas may be in very short supply at any price. At that time the competition may be among sucrose, cellulose, and coal. This paper addresses the cost of ethanol production in some detail and takes a brief look at acetone, butanol, and glycerin from hydrocarbons and sucrose, since these are cases where significant volumes of sucrose might be utilized.

Before a comparison of hydrocarbons and sucrose for chemical use is made, a look is in order at the alternate uses of sucrose. There is no point looking at markets for sucrose as a chemical feedstock even at a low cost if its use as food or animal feed commands a better price. The comparison also should include fuel as well as chemical uses of the products made from sucrose, since this use also is influenced by the increased prices for hydrocarbons. In this comparison the value of the material to the user must be considered. As a start, let us compare food and fuel values. Sugar at 20¢/lb in the supermarket sells for $28/million Btu. Gasoline at 62¢/gal (including taxes) is equivalent to $5/million Btu. The higher value for the sugar can be attributed to the fact that it has extra value, namely nutritional, preservative, and flavor values. Thus, until all food markets for sugar are satisfied at all prices down to the fuel value, no sug-

ar will be offered to produce fuel at the fuel-valued
price.   Of course, once the food market at the current
price is satisfied and prices drop, other markets may
be able to use sugar before the price reaches the fuel
price.   Another large market that can afford to pay a
nutritional premium is the animal feed industry, which
now is a large user of molasses.   Thus, there is a
hierarchy of needs and willingness to pay, in the order
food, feed, then fuels.   Further consideration of
other markets leads to the extended hierarchy shown in
Table I (with apologies for the forced alliteration).

Table I.   Hierarchy of Farm Products Uses

1. Farmaceutical	10. Framing and lumber
2. Firewater	11. Feedstock for chemicals
3. Fragrance	12. Fertilizer
4. Flavorant	13. Friability aid, erosion
5. Functional aid for food	control, etc.
6. Food	14. Fuel
Flesh, fowl, fish	For family fliver
Fats and oils	Flying
Fillers	Family furnace
7. Feed	Factory and electri-
8. Fiber	city generation
9. Film and sheet	

Intermediate between feed and fuel uses are those that
utilize the one, two, and three dimensional properties
of cane fiber or sucrose derivatives, for example,
rayon or polyethylene, fiber, paper, and Celotex
board.   The chemical and bulk physical properties are
reflected in uses as a chemical feedstock, fertilizer,
and aid to improve the quality of the soil, prevent
erosion, and hold moisture in the soil.   These all
have a greater value per lb than just using the chemical bonding energy released on combustion.   Also,
there are values higher than the nutritive value of
food.   These are related to higher physiological activities of the material such as, use as a drug or
medicine, flavor, or fragrance.   The beverage use of
alcohol, tax paid at retail, commands at least $40/gal,
which is equivalent to $60/gal for gasoline.   Thus,
sucrose as a sweetener is worth more per lb than corn
starch, which provides calories but no flavor.   Functional aids, that is ingredients which improve the
properties of food, shelf life, or preparation ease
have a premium.   Sucrose as a preservative in marmalade
is worth more than a food that only provides calories.

It should be noted that the higher value products usually are sold in smaller volumes than the lower valued products. For example, the annual per capita consumption in the United States is 90-100 lb of sugar versus 2700 lb of gasoline. Economic incentives lead the producer to fill the highest value market even at low volumes, then to "spillover" to the next lower price, larger volume market. Frequently, product differentiation is used to avoid loss of the premium price in the first market while gaining sales volume in the second. For example, alcohol is sold in beverage (taxable and tax free) and denatured grades.

Within each category there also is a hierarchy. Meat is worth more than fats and oils, which in turn are worth more than starchy foods, which provide merely calories. In the fuels category, transportation fuel and home heating have higher values than fuel for running factories or generating electricity.

Occasionally, there is an inversion of the order, but only in unusual circumstances, for example, the snowbound cabin dweller burning his furniture for warmth. For chemical uses of sucrose to rank higher than food and feed would require quite a revolution in human values. On the other hand, at the food and feed price for sucrose, economic incentives would be more likely to lead to the use of sucrose as a chemical intermediate than as a source of fuel.

Looking further at chemical opportunities, a search was made for large volume chemicals that could afford the cost of sucrose as a raw material. When the volume of organic chemical production is graphed against price per lb for 150 leading organic chemicals (2) in a log-log plot, a nearly straight band is found as shown in Figure 1. However, if the type of chemical is considered, the display no longer is a band but a series of overlapping patches, as shown in Figure 2. Here, a hierarchy can be seen, similar to that described previously. Most primary petrochemicals are in the high volume-low price category. Chemicals with macromolecular structure, that is, plastics, resins, and elastomers, are in the medium-to-large volume and at a higher average price than basic petrochemicals. Materials with physiological activities such as medicinal chemicals, flavorants, and some pesticides have low volumes and high values. Chemicals with other functional values, such as rubber processing chemicals, plasticizers, and surface active agents, sell in moderate volumes at prices between the two extremes.

With the idea of utilizing large amounts of sucrose and releasing large amounts of petroleum for

A. Crude Products from Petroleum, Natural Gas
B. Cyclic Intermediates
C. Dyes and Pigments
D. Medicinal Chemicals
E. Plastics and Resins
F. Elastomers
G. Plasticizers
H. Surface Active Agents
I. Pesticides
J. Miscellaneous Organic Chemicals
K. Flavorants
L. Rubber Processing Chemicals

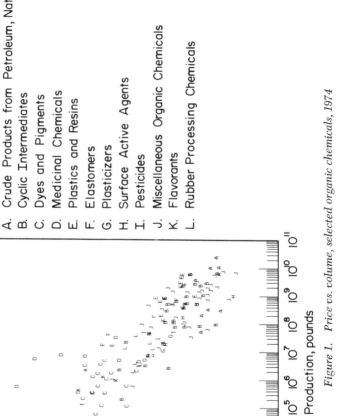

Figure 1. Price vs. volume, selected organic chemicals, 1974

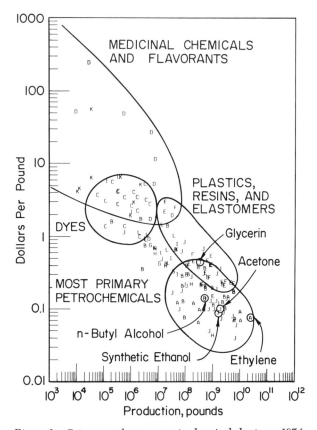

Figure 2.   Price vs. volume, organic chemicals by type, 1974

other uses, the feasibility of producing ethanol, butanol,acetone, and glycerin were chosen for further analysis.

Industrial Ethanol

        Until the petrochemical and natural gas boom, in-
dustrial ethanol was manufactured primarily by fermentation of molasses.   The production of ethylene with
a selling price of only 3¢/lb combined with increased
use of molasses in animal feed led in the United States
to the manufacture of ethanol primarily by hydration of
ethylene.
        The current market for industrial ethanol exceeds
700,000 tons/year (3).   Additional ethanol is used
captively by some producers to manufacture acetic

acid and acetaldehyde, thus leading to a still larger production than the sales figures directly attributed to ethanol. The selling price of industrial ethanol has risen rapidly in the last few years, reflecting ethylene price increases from the range of 3¢ to 4¢/lb to 10¢ to 12¢/lb. The present selling price of ethanol is approximately $1.15/gal, of which the ethylene cost is approximately 50¢ and the non-raw material costs total 65¢/gal. The non-raw material costs include marketing charges and profit in a business that has many small customers with specialized requirements.

Ethanol from Cane Juice. Research conducted by Battelle Columbus Laboratories for the Fuels from Biomass Program Office of the Division of Solar Energy of the US Energy Research and Development Administration (ERDA) has focused on the possible manufacture of ethanol from sugar crops. The prospects for manufacture of ethylene from fermentation ethanol is included in this research project. Ethylene production from sugar crops is considered a long-range opportunity because the price of petroleum would have to nearly triple before it would begin to appear economically attractive. Accordingly, ethylene from fermentation ethanol is not discussed in this paper.

The consolidated manufacturing costs for an ethanol fermentation facility that is capable of making about 70 million gal per year of 190 proof alcohol is shown in Table II. The major cost element in fermentation ethanol is the sugar solution that is the raw material. Juices that could be extracted from sugarcane, sugar beets, and sweet sorghum were investigated by Battelle. Sugarcane juice appears to be most promising for large-scale operations because the co-operational bagasse can be used to achieve fuel self-sufficiency. For smaller scale operations, by-product molasses may be as good or better but the price volatility of this by-product could be a source of concern. Battelle estimates that close to 15 lb of fermentable solids would be required to make one gal ethanol. A reasonably optimistic cane juice cost is approximately 6¢/lb (fermentable solids basis). Therefore, the sugars alone would cost about 90¢/gal of ethanol.

The major capital cost is for a large number of fermentation tanks. Therefore, smaller plants may have nearly as low capital cost per gal because one simply uses fewer, large-size fermentation tanks. The major operating cost is the steam required to distill the ethanol and to perform various cane juice concen-

Table II.

Consolidated Manufacturing Costs - Ethanol Plant
(In millions of 1976 dollars)

Basis: 70 million gallons of 95 percent ethanol and 224,400 tons
dried stillage

Annualized Capital Charges	14.9 [a]
Estimated Operating Costs including denaturant	22.2
Sugarcane Juice	55.8
TOTAL	92.9
Stillage By-Product Credit	(11.2)
NET MANUFACTURING COST	81.8
NET MANUFACTURING COST PER GALLON	1.17

(a) Based on capital investment of $127 million, 60% debt, 14% return
after taxes.

tration operations. The availability of cheap bagasse
to accomplish this steam generation is a definite advantage for sugarcane. However, beet juice or molasses may be acceptable if a supply of cheap steam can be
assured. Another relatively high cost item is working
capital, because concentrated cane juice must be stored
during the off season and considerable ethanol inventories are required to service the market.
Stillage, that is, still residue and spent yeast,
is a nutritious animal feed concentrate. The stillage
from this ethanol process does not have a composition
as desirable as that obtained from grain ethanol operations. Protein content is low and salts content is
high. This stillage was estimated to be half the
value of the conventional grain-based products.
The net costs are calculated to be about $1.17/gal.
These costs include a charge for a return on equity of
14% after taxes, assuming that 60% of the capital for
the venture were borrowed at 8.75% interest. Marketing charges and other promotional expenses that could
add approximately 35¢/gal are not included. If more
conservative financial policies were used (e.g., 30%
debt and 15% after taxes return on equity), the manufacturing cost would rise to about $1.30/gal and the

selling price might be close to $1.70.

These preliminary calculations indicate that ethanol by fermentation of sugarcane juice is close to equivalence with ethanol derived from petroleum or natural gas liquids at $1.15/gal. As described above, the competitiveness of sugarcane with the nonrenewable resources depends heavily on the cost of the fermentable sugars. If one had to purchase molasses on the open market, the venture could be in deep trouble during times of high molasses prices, occasioned, perhaps, by a corn crop failure.

The value of the stillage is another critical factor. If this product could be upgraded to sell more competitively with other distillers dried solubles, the economics of this ethanol process would improve but, if the product proved unacceptable, a corresponding depressing effect would be observed.

Another sensitive element is steam, the cost of which is minimized by employing bagasse as the raw material. When new vacuum fermentation processes, such as that under investigation by the University of California (4), reach full development, steam will be used more efficiently and the complexity of the ethanol facility and capital cost could be reduced significantly.

Ethanol from Petroleum. At what relative price will sugarcane replace petroleum or natural gas liquids in the manufacture of industrial ethanol? A sophisticated answer to this question cannot be given in purely economic terms because many business considerations are involved. For example, the structure of the chemical industry is such that marketing relations are of great importance. In addition, location and distribution considerations may be of considerable importance. A company may already have a well-amortized facility to manufacture industrial ethanol from a nonrenewable resource. Therefore, the switch to a renewable resource would carry a penalty. Beyond the business considerations, there are such considerations as maintaining levels of employment at established locations. Despite noneconomic considerations, it is important to have the economic crossover points in mind because they are the standards against which modified policies are compared.

Construction of a simple equivalence diagram is shown in Figure 3. 4.2 lb of ethylene are required per gal of ethanol manufactured by petrochemical processes. The cost of ethylene recently has been approximately 0.010 times the cost of petroleum, when the petroleum is expressed in dollars per bbl and ethylene

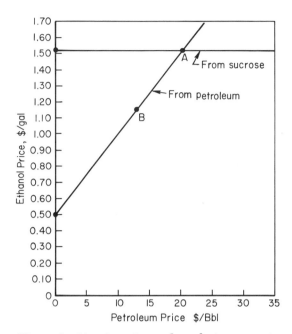

Figure 3. Manufacturing and marketing costs, including profit, for ethanol

in dollars per lb. Using this relation, the ethanol
selling price (including marketing profits and costs)
can be plotted on the Y axis versus the petroleum selling price as on the X axis (see Figure 3). This
straight line can be compared with possible fermenta-
tion-ethanol selling price. For example, adding a
30% marketing charge to the $1.17 manufacturing cost
in Table 2, yields the horizontal line at $1.52, shown
in Figure 3. The intercept of the petrochemical-etha-
nol line with fermentation line occurs at approximately
$19/bbl. This equivalence point (A) can be compared
with point B, which indicates that at the current petroleum selling price of approximately $13/bbl, the corres-
ponding ethanol selling price is $1.15/gal. At the
current price of petroleum the manufacturing cost for
ethanol would have to be 88¢ to allow for marketing
charges of 30%. This would correspond to a cane juice
price of 2.5¢/lb of solids, less than half the current
cost.

n-Butyl Alcohol and Acetone

Acetone is an important intermediate in the synthesis of methyl methacrylate and bisphenol-A, both of

which are used in making plastics, and in the synthesis
of methyl isobutyl ketone, methyl isobutyl carbinol and
isophorone, which are used as solvents. In addition,
acetone is used directly as a solvent for cellulose
acetate spinning and for protective coatings. In 1975,
production of 796,000 tons was reported. The major
use of n-butyl alcohol is as a solvent and intermediate
for making butyl acetate and glycol esters for solvent
use. It is also used in making certain plasticizers,
amine resins and butyl acrylate and methacrylate resins. Production in 1974 was 278,000 tons.

During World War I, a process for fermentation of
molasses to give n-butyl alcohol, acetone, and minor
amounts of ethyl alcohol was developed by Chaim Weis-
mann. One lb of molasses is reported to yield 0.25 lb
butyl alcohol and 0.10 lb acetone (5). As was the
case with ethylene, cheap natural gas liquids as a
feedstock led to abandonment of fermentation for production of these chemicals.

Acetone now is made from propylene by two processes. In the first, propylene is hydrated to isopropyl
alcohol, which is then oxidized to acetone. In practice, 1.1 lb propylene yields 1.2 lb isopropyl alcohol,
which in turn gives 1 lb acetone. Most new facilities
use a second process, namely coproduction of acetone
and phenol from cumene, which is made from propylene
plus benzene.

Butyl alcohol can be made from 2 molecules of
ethylene via acetaldehyde, aldol, and crotonaldehyde.
One lb butyl alcohol requires 1.1 lb ethylene. Alternatively, propylene can react with carbon monoxide
and hydrogen to give a 4:1 mixture of n-butyl alcohol
and isobutyl alcohol. One lb of alcohol requires 0.58
lb propylene.

For calculation purposes, butyl alcohol from ethylene and acetone from propylene via isopropyl alcohol
are assumed. One lb of butyl alcohol and 0.4 lb acetone from cane juice at 6¢/lb of soluble fermentables
gives a combined manufacturing and marketing cost and
profit of 36.6¢, assuming operating and capital costs
are twice that for ethanol since the fermentation is
more difficult to control, and that the stillage is
worth half the value of grain stillage. There are
many vitamins in the stillage, making it valuable for
non-ruminant feed so that this is conservative.

At current prices, 1 lb butyl alcohol sells for
22¢ and 0.4 lb acetone for 6¢ or a total of 28¢.
With a current price of ethylene of 12¢ and propylene
at 8.4¢/lb, and assuming that propylene remains 0.7
times the price of ethylene and that the price of ethy-

lene per lb is 0.12, it would be economic to produce
n-butyl alcohol and acetone from sucrose when petroleum costs $17/bbl.   This assumes a marketing profit
and cost of 30% over manufacturing profit and cost.

Glycerin

About half of all glycerin is used directly in
drugs and cosmetics, tobacco, and foods and beverages
as a humectant, functional aid, and vehicle.   Chemical
derivatives from glycerin include alkyd resins, polyether polyols, and explosives.   It also is used as a
plasticizer in cellophane.   U.S. production in 1974
was 175,000 tons, about 60% of which was synthetic.
Three different processes are used to make glycerin
synthetically, all starting from propylene.   Each process is fairly complex, involving 3 or more separate
synthetic steps (6).   Natural glycerin is obtained
from fats as a byproduct of soap making and fatty acid
production.

Glycerin can be produced from sucrose by fermentation (7) or by hydrogenation.   The latter reaction,
considered here, yields four molecules of glycerin for
each molecule of sucrose from high test molasses, that
is, concentrated cane juice.   A plant using this technology of 15 million lb per year capacity was operated
at New Castle, Delaware, until 1969 by Atlas Powder
Company (now ICI-United States, Inc.).

A quick look at the economics in today's market is
of interest.   In 1960 when the price of quota-exempt
sugar was 3¢/lb and shipping and insurance from Cuba
0.5¢/lb, the price for glycerin was 29¢/lb.   If the
yield was 80% and no by-products were sold, operating
costs, capital charges, and manufacturing profit were
17¢ (assuming marketing profit and cost was 30% of the
total manufacturing profits and costs).   In 1969 when
the plant closed down, presumably because of the high
cost of sugar, 8.8¢/lb, and low price for glycerin,
24¢/lb, operating costs, marketing costs, and profits,
if any, were 14¢.

Cane juice costs about 6¢ lb and raw sugar 12¢/lb
(if reasonable profit is obtained).   Assuming that
high-test molasses of the desired concentration is
available for 8¢/lb of solids, and that operating costs,
capital charges, and manufacturing are double the 1960
figure, that is, 34¢, a cost of 44¢/lb before marketing
costs and profit would be incurred.   With marketing
costs and profits, the figure is 57¢/lb.

Since 0.625 lb of propylene is required to produce
one lb of glycerin, and assuming 30% marketing profit

and costs, the price of petroleum would have to rise to about $35/barrel for this to be an economic alternative.
   If modern technology could reduce the cost, or if larger capacity than the 15 million lb per year for the old plant was used, say 100 million lb per year, the non-raw material manufacturing should be reduced 24.5%. (This assumes a scale factor of 0.85, that is, cost ratio = (size ratio)$^{0.85}$). This, plus an improvement in yield to 90%, would give costs of 46¢, very close to the current 48¢/lb. This calculation is not meant to be precise--it is rather an indicator that this chemical may be promising now or in the near future.

Agricultural Considerations

   The sugar industry may choose to manufacture ethanol, butanol, and acetone from by-product molasses or use cane juice from sugarcane dedicated specifically to this product. The selling price of by-product molasses is subject to wide swings, depending on the availability of alternative carbohydrate sources for ruminant nutrition. This route could be quite appropriate for a sugar company with a sizable enough molasses operation so that it need not buy molasses from outsiders.
   Alternatively, one could envision sugarcane chemical farms in which the cane juice would be dedicated to chemical production and facilities for crystallization sucrose might not be installed. Such farms probably would be needed for the production of large quantities of ethanol, such as the venture described in this paper. Accordingly, some information is provided on the likely cost of growing sugarcane.
   The capital cost of a 640-acre farm in Florida totals about $1.6 million. The necessity for growing seed cane and leaving some of the land fallow contributes to the fact that only approximately three quarters of the land is producing for shipment to the central processing facility in a given year.Sugarcane production costs in several typical areas are shown in Table III. Only a part of the millable cane is fermentable sugars that can be used to manufacture ethanol. The costs,in terms of fermentable solids,are shown in Table IV for sugarcane and other sugar crops. When the cost of extracting the cane juice is added, all regions appear to have a cost of fermentable sugars of approximately 6¢/lb of solids.
   The central processing facility to manufacture ethanol on 70 million-gal/year scale might serve an area of 50,000 to 100,000 acres. In locations where

Table III. ESTIMATED 1976 SUGARCANE PRODUCTION COSTS, MAINLAND UNITED STATES

	Texas		Louisiana		Florida, Muck Soil		Florida, Peat Soil		Florida, Sandy Soil	
	$/Acre	$/Ton	$/Acre	$/Ton	$/Acre	$/Ton	$/Acre	$/Ton	$/Acre	$/Ton
Tons millable cane/acre	38		25		42		35		31	
Preharvest costs										
Seed	12.00	0.32	16.48	0.66	16.76	0.40	16.76	0.48	16.75	0.54
Fertilizer	41.35	1.09	35.05	1.44	29.19	0.70	34.82	0.99	90.97	2.93
Chemicals	41.25	1.09	26.30	1.05	32.00	0.76	28.31	0.81	28.82	0.93
Labor	38.61	1.02	53.26	2.13	45.27	1.03	53.46	1.53	55.60	1.79
Fuel and lubricants	6.44	0.17	10.40	0.42	12.04	0.29	12.04	0.34	13.77	0.44
Repairs	9.96	0.26	17.83	0.71	13.86	0.33	13.86	0.40	15.30	0.49
Interest on operating capital	11.85	0.31	6.93	0.28	7.31	0.17	7.75	0.22	12.49	0.40
Miscellaneous	17.00	0.45	7.28	0.29	13.72	0.33	13.57	0.39	13.63	0.44
Subtotal	178.46	4.71	174.53	6.98	170.15	4.05	180.57	5.16	247.34	7.96
Harvest costs										
Labor	34.45	0.91	34.52	1.36	115.62	2.75	101.16	2.89	64.11	2.07
Hauling	76.00	2.00	30.00	1.20	50.40	1.20	42.00	1.20	37.20	1.20
Fuel and lubricants	20.60	0.54	9.34	0.37	17.30	0.41	17.30	0.49	17.30	0.55
Repairs	38.52	1.01	31.00	1.24	26.21	0.62	26.21	0.75	26.21	0.85
Interest on operating capital	3.54	0.09	0.85	0.03	10.60	0.25	9.07	0.26	8.44	0.27
Miscellaneous	--	--	--	--	14.37	0.34	14.37	0.41	14.37	0.46
Subtotal	173.11	4.55	105.71	4.22	234.50	5.57	210.11	6.00	167.63	5.40
Machinery ownership costs	57.54	1.51	54.21	2.17	47.57	1.13	47.57	1.36	48.99	1.58
Land charge	90.00	2.37	75.00	3.00	140.00	3.33	115.00	3.29	105.00	3.39
Management charge @ 7% of gross receipts	39.27	1.03	28.84	1.15	49.07	1.17	40.88	1.17	36.21	1.17
TOTAL COSTS	538.38	14.17	438.29	17.53	641.29	15.26	594.13	16.98	605.17	19.52

Source: Battelle-Columbus estimates based on unofficial U.S. Department of Agriculture crop budgets and Texas A&M University crop budgets (adjusted to 1976 price levels where necessary). The dry solids content of millable cane is approximately 27 percent.

TABLE IV. COMPARATIVE RAW MATERIALS COSTS OF SUGARCANE, SUGAR BEETS, AND SWEET SORGHUM

| | Cents per Pound | |
	Fermentable Solids	Combustible Organics
Sugarcane		
Texas	4.6	0.9
Louisiana	5.4	1.2
Florida, muck soil	4.2	1.3
Florida, peat soil	4.7	1.4
Florida, sandy soil	5.3	1.6
Hawaii	4.2	1.3
Sugar Beets		
California Coastal Region	4.6	1.0
Texas High Plains	5.2	1.4
Southern Minnesota	6.0	1.9
Sweet Sorghum		
Syrup varieties	3.9	1.1
Sugar varieties	4.3	1.5

Source: Battelle's Columbus Laboratories estimates.

the grinding season is relatively short (e.g., Louisiana), sweet sorghum might be grown to extend the season by harvesting this crop during the several months that precede the sugarcane harvest. The cane dedicated to chemical use does not need to be processed as quickly as that processed for sucrose since frost reduces sucrose content but does not affect total fermentable solids to any great extent. This means that the grinding season can be longer in regions with fall frost, which improves the economics by allowing a smaller scale mill or the sharing of a mill that now produces sucrose.

Abstract

Although sucrose technically is suited for making a large variety of chemicals, ethyl alcohol, n-butyl alcohol, acetone, and glycerin are the ones with large

volume markets.   They can be made directly from cane
juice or high test molasses,saving the cost of crystalization and refining.
     This brief analysis shows that glycerin may be
economically attractive right now, if made in a large
size plant.   At $17/barrel of oil, it should become
feasible to make n-butyl alcohol and acetone by fermentation of sucrose.   If oil prices rise to $19/barrel,
industrial ethanol becomes attractive.
     Thus, for these chemicals that can be made direct-
ly from sucrose,opportunities exist that make a detailed feasibility study a worthwhile, next step in the
commercialization process.

Literature Cited.

1. Kollonitsch, Valerie, "Sucrose Chemicals", The
   International Sugar Research Foundation, Inc.
   Bethesda, Md. U.S.A., 1970.
2. Synthetic Organic Chemicals, 1974, U.S. International Trade Commission.
3. Chem. & Eng. News, (1976), June 7.
4. Cysewski, G.R., and Wilke, C.R., "Fermentation
   Kinetics and Process Economics For the Production of Ethanol, filed as a PhD thesis (March,
   1976).   Prepared for the U.S. Energy Research
   and Development Administration Under Contract
   W-7405-ENG-48, University of California, Law-
   rence Berkeley Laboratory.
5. Lowenheim, F.A., and Moran, M.K.,"Butyl alcohol" in
   Faith, Keyes and Clark's Industrial Chemicals,
   4th ed, Wiley, 1975, pp 178-185.
6. Lowenheim, F.A., and Moran, M.K., "Glycerin" in
   Faith, Keyes and Clark's Industrial Chemicals,
   4th ed, Wiley, 1975, pp 430-441.
7. Megna, John, "Utilization of Sugar in Industrial
   Fermentation", ISRF Study No. 312, 1971-72,
   unpublished.

Biographic Notes

     William J. Sheppard, Ph.D., Senior Res. - Energy
Systems and Environmental Research Sect.   Educated at
Oberlin Coll. and Harvard Univ.   A technical economist
with a background in organic chemistry.   Battelle Me-
morial Institute, Columbus Laboratories, 505 King St.,
Columbus, Ohio 43201 U.S.A.

An Outsider's View of Sucrochemistry

B. J. LUBEROFF

Consultant, The Bassett Bldg., Summit, N.J. 07901

In this symposium I am billed as the new kid on the block, and I surely am. Preparing this paper has been a truly educational experience for me, and now I am an expert. I define an expert as anybody who heard about the problem a half hour before I did.

I assume that my auditors are a slightly heterogeneous group with a preponderance of research people from the sugar industry and from academia. I think this composition is unfortunate because the really fruitful discussions do not occur among those concerned with the intricacies of sugar chemistry talking among themselves. This is so because the problem is not the intricacies of sugar chemistry. Rather, the challenge, as it has been given to me to discuss, is "how can more sugar be brought into channels of commerce?".

To answer this, I shall pretend that I am the Director of New Ventures for the Amazing Sugar Company, and share with you, some of my thoughts as I play that role.

Like Lincoln, I first like to ask,"Where are we?", before asking "Whither?". Put otherwise, before doing R & D, find out what business you are in. Then try to find out what business you would like to be in. According to John Hickson at the International Sugar Research Foundation, 98% of all sugar sold is used to sweeten something. I verified this by consulting Chemical Market Abstracts back through 1973 and found only one reference to sugar used for any other purpose; that was making detergents and the only process cited was a Japanese one. So, it seems fair to say that the business of the sugar industry today,as it always has been, is selling sweetness. Table I depicts the magnitude of the business.

Table I.   World Sugar Production and Consumption
                   (Millions of Metric Tons)

Crop Year   1975/1976*   1974/1975

Starting Stock   16.4   16.1
Production   82.0   79.1
Total Supply   98.4   95.2
Consumption   80.8   78.8
Carryover   17.6   16.4

*Estimated

     Now, what business do we want to be in?   Trying
to answer that, brought to mind a conversation I had
back in the early part of this decade when the chemical
industries were in more of a state of panic than they
are now.   I was at one of those sequestered conferences
they hold in the New England woods.   Our topic was
Long Range Planning.   After two days I became concerned that we had not defined the time frame of "long
range" and expressed this concern to one of the other
conferees.   He said, "Oh, that's easy.   In long range
planning, the time frame is inversely proportional to
how dismally you view the future.   Right now we're
concerned about tomorrow's survival."

Table II   Major Sweetener Market Estimates (million lbs.)

Date	Honey, Misc.	Dextrose	Corn Syrup	HFCS	Saccharin	Sugar	Total
1975	200	1,081	4,071	950	1,484	21,303	29,089
1981	200	1,390	5,216	5,000	1,863	21,000	34,669
1986	200	1,610	6,046	8,000	2,377	20,820	39,053

Source: Business Communications Co., Inc., Stamford, Connecti-
           cut, USA.   Press Release, Study C-005, "Sugar Substi-
           tutes and Artificial Sweeteners", published May, 1975.

     Concern over the survival of the current business
is always there.   If you have a profitable business,
you can rest assured that somebody is going to attack
it, either by selling the same product or by selling
something that does the same job.   Those of you who
have seen synthetic sweeteners come and go know exactly
what I mean.   At Amazing Sugar we have found the big-
gest attack on our sweetness business comes from corn-
derived sweeteners (Table II) so we got into fructose.
(Now we are not so sure that was a good idea.)   We are
not making synthetic sweeteners, simply because we like

to make our own executive decisions rather than have
FDA make them for us.   Of course, we now may have to
go part way and chlorinate some sucrose, but our current business seems to be holding up without much product R & D.   The problem with focussing R & D on doing what you are already doing is that almost by defin-
ition, the results you can expect are limited.   We
should do some defensive research, but at this point I
have no reason to feel that at Amazing Sugar we are not
doing enough.

   Returning to "what business do we want to be in?",
I am reminded of another conversation I had, this time
with a neighbor.   He was an attorney and I asked him
in what area of the law he concentrated.   "Commercial
Law," he replied.   "What is that" said I.   "Anything-
I can make a buck at," he exclaimed.   From what little
I know of past R & D in sucrose chemistry, I get the
feeling that its objectives are just about that dif-
fuse.   I do not think that is a good idea.   We pick
at every hydroxyl group with every reagent we can think
of.   We have kept too many options open.   We have
to get some focus.   The question is where do we find
the focus?   I think that one of our difficulties is
that some of us have tried to define our problem as
"Let us use sucrochemistry to increase the sales of
sucrose".   Thus, we have built a fallacy into the
whole argument.

   Let me explain by looking closely at the sheer
size of the sugar industry.   It is BIG.   The dollar
volume of the world sugar trade is exceeded only by
that in petroleum, grain and coffee.   In the U.S., over
20 billion pounds of sugar is sold a year.   As Henry
Hass has noted, that is bigger even than ethylene.
This is important because, if any new product is going
to make even a ripple,say 1%,in sugar use, we are talking about perhaps 400 million lb/yr of, for example, a
sugar detergent.   That just about equals the total
annual tonnage of alkyl benzene detergents sold in the
U.S. today.   Another way to say it is if we found a
product that used even a tenth percent of total U.S.
sugar production it would represent a raw material cost
of about $3 million a year and a product sale of close
to $10 million a year.   That is no greater than the
total annual business of a lot of thriving, publicly
owned chemical companies.

   I submit, therefore that, thinking of sucrochemistry as a means to sell sugar, is unreasonable.

   I prefer instead to identify a market.   Learn all
I can about it.   Then go after it monomaniacally.

R & D is paid for only because it is expected to
satisfy a mass human need in the market place, yet most
of us do not bother to find out much about markets.
R & D is not done to sell sugar, and it is also not
done to satisfy our personal curiosity.

So let us focus our attention on some human needs.
There are a few real needs, and there are the manufactured ones. Whether the market is real or synthetic,
you must first convince yourself that the market is
there and is big enough to justify your effort. Then
you must convince some bill-payer that you are right.

Basically there are two ways to identify a market.
One, is to go out and see what the market has been doing. What do people buy? What do they pay for it?
What are they using it for? What alternatives are
there? Another way is to give them something they
never even thought about. Create a market.

First, if you are going to create a market there
is no point in going out and seeing what the market is
doing. Market surveys are expensive and dubious, at
best. Instead, look to yourself and see what kind of
need you have and make the assumption that everybody
else has that same need. This is not a new concept;
you will find it in the Bible, where it says, "Do unto
others what you would have them do unto you".

The Pep Pill

I tried this approach when I was thinking about
sucrochemistry and this is what happened. I said to
myself, "Self, what do you need?" Self answered,
"Energy!" I do not mean global energy, I mean Pep.
We all need Pep. Being a chemist, I know that Pep
comes from carbohydrates and 20% of our carbohydrates
are eaten in the form of sugar. So, why not eat a
candy bar? No, that is self-indulgent, hedonistic!
The work ethic does not let me eat candy bars, they
taste too good. Instead, being modern, I would find
more acceptable some sort of pill, a Pep Pill. Let
us invent the hard, compact, portable, sugar pill.
We will coat it so it will not taste good. Then, we
will put it in a drugstore-type, plain wrapper, call
it something sexy like α-D-glucopyranosyl-β-D-fructo-
furanoside, label it as all "natural and organic", and
price it high so that it has "class". In addition to
all the attributes we have outlined,this pill will have
a wonderful placebo effect.

We now have a new project, what do we do next?
We write down all the things that could go wrong.
These usually break into two categories, fatal flaws,

yes/no, go/no-go parameters; and those that can be
represented as quantitative continua.   Among the go/no-
go things, are how FDA will react to this new product.
We are probably safe there; GRAS list all the way, and
I doubt that there is a patent block,   Although I am
sure you can think of other candidate death sentences,
note that those we just listed are in the area of The
Laws of the Land, not the Laws of Chemistry.

If you cannot think of any fatal flaws, the next
step is the quantification step.   Usually, that is
based on some measure of profitability such as dis-
counted cash flow or return on investment.   Investment
is the cost of the factory, in this case, a simple one
- a tabletting machine and a packaging line.   We could
even rent the use of them without having to make a fixed capital investment.   Energy requirements? Trivial.
Labor costs?   Not much at all.   Raw materials?   No
problem.   We know the yield and the cost of sugar ...
I think.

It sounds like we really ought to proceed.   But,
how much do we make?   Remember - look to the market,
how much is being sold?   None - unless you include
Life Savers.   How much can we sell?   Let us see: two
pills a day at a nickle each, X 200 million tired
Americans, and 1% market penetration equals, $40 mil-
lion/yr.   Too much; how about a tenth percent market
penetration?   Guessing right is where the risk comes
in.

We could go on like this, but I think you get the
message.   You must keep shuttling back and forth between the market and the technology.   In this particular case I chose a product that has practically no
technology, no patents, no FDA, so you can see exactly
what THE problem is.   In this case THE problem would
be a very good advertising campaign.   Did you know
that a single cigarette company can spend $30 million
/yr in magazine ads alone?   Since we are paid by the
market, we have to abide by its dictates.   We have to
give the public what it wants, or create that want.
Technologists, even scientists, have to be cognizant of
this and work as close to the market as they dare.
Those who work really close to the market are called
Product Champions or, if they set up in business for
themselves, entrepreneurs.

The Product Champion and Exclusion Chart

Good product champions are extremely valuable.
They shuttle between the market and the laboratory.
In the market they will look at dollar volume and unit

cost.   One nice technique for doing that is the exclusion chart invented by Herman Zabel and kept alive by
Roger Williams.   Figure 1 is a representative exclusion chart.   What these charts say is that in a given
market area, plasticizers (in Figure 1) for example,
could in 1970 achieve a volume of over 6 million lb/yr
only if the unit price were below 28¢/lb.   If you plot
the logarithm of volume vs unit price in a given cata-
gory of materials, you get the kind of L-shaped point
array that is called the Exclusion Chart.   Once in a
while,you may find a product in the upper right,exclud-
ed area.   When you do, you can be sure that it is
there because it has very unusual properties.   Teflon
is one such.

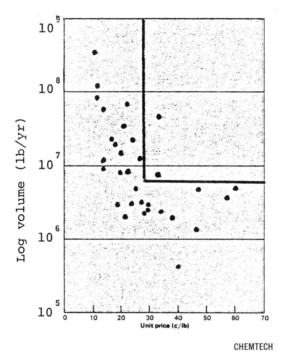

CHEMTECH

Figure 1.   Exclusion chart for plasticizers, 1970 data
(Source: Roger Williams, CHEMTECH, Oct. 1973,
593)

Note: 1970 data, after Roger Williams CHEMTECH,
      Oct. 593, 1973.

    This technique will help you decide which price/
volume relationship you want to go after.   Do you want

a small volume specialty product that commands a high
unit price? Or,are you shooting for commodity volumes?
One way you can decide is to look again at sugar.
With sugar at 15¢/lb, each carbon atom is going to cost
you about 30¢/lb and sell for perhaps 60¢/lb.   That is
about twice what the carbon atoms cost in basic petrochemicals, at present.   As the source for petrochemicals is depleted, the price ratio will reverse because
sugar is a renewable resource.   On the other hand, you
may be anxious to get into business now.   With a new
product, "now" means five years <u>minimum</u> from concept to
sales.   Thus what you will have to do is find something that is able to command 60¢/lb for carbon, or
else use the oxygen that is in sucrose for some useful
purpose.   The rule of thumb I have used, is that every
unit process, followed by a unit operation, that is,
every chemical conversion, plus its following purification step,costs as much as the raw materials fed to it.
Thus, if you feed sucrose into a process with something
else that also is worth 15¢/lb, use all of it in your
product, and then purify what is made, you are talking
approximately, about a 30¢/lb product.   That could in
1970, be a high volume plasticizer if the yields are
right and the rule of thumb is not too far off.

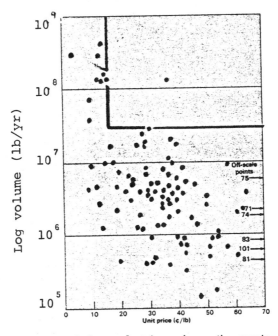

Figure 2. Exclusion chart for surface active agents,
1970 data

It seems to me that we have to do two things: one
is to find some class of product that is selling in a
volume area, of interest, for between 25 and 50¢/lb.
The other, is to make sure that the kind of use presents advantages to materials that have a lot of oxygen
in them.   How about synthetic detergents?   Figure 2
indicates that in 1970 the critical price was 15¢/lb.
We just have decided that we cannot make a synthetic
detergent out of sugar for 15¢/lb.   But what is the
critical Exclusion Price for detergents today?   Maybe, after we look, we will decide to be satisfied with
a relatively small volume, specialty product.   Maybe
not!   Are we looking at surface active agents for
wash day use or in specialty uses?   Do we know those
markets?   What might some other products be?

One cannot go through this sort of exercise without
coming up with a few candidate ideas and I had some.
Some, maybe all, may be old.   I have not done any sort
of literature search, but, I excuse that by pointing
out that old is not necessarily bad.   Look what is
happening to coal and windmills.   (Look what my wife
pays for antiques.)   So here are a few, market-direc-
ted ideas on sucrochemicals that occurred to me.
How about:
- Phosphate esters for plasticizers.
  If enough phosphorus is added they
  will be polymer compatable and also
  impart flame proofing properties.
- Bromo --- woops!   That's too expensive
  --- O.K., chloro esters for the same
  purpose.
- How about mono-acrylates that could be
  knit into polymers as dye sites, anti-
  statics, solubilizers and flocculating
  agents.
- Then there are fiber intermediates.
  That is a growing market so let us take
  a closer look at an exemplary project
  there.

Polyester fibers have been one of the fastest
growing segments of the chemical industry.   It is well
known that on hot days, synthetics, in general, could
be more comfortable, i.e., hydrophilic, like cotton.
Oxygen is hydrophilic and sugar has lots of that.   Can
a fiber be made out of sugar?   I thought that maybe
one could,when I learned that sucrose could be converted to hydroxymethyl furfural.   The kind of chemistry
that occurs to someone with a market orientation is in
Figure 3.   As can be seen, one has three diols, the

unsaturated one and both the cis and trans versions of
the saturated one.   They could be rather fascinating
things.   Not only could they offer a little extra
hydrophilicity but they also have a certain kind of
rigidity imparted by the cyclic structure, that could
be interesting in a fiber.   Going down the left side
of the Figure you will see that you can also make di-
carboxylic acids.   What could be nicer?   Here is a
polyester, all made from sugar.   On the right side is
the intermediate product, the dialdehydes.   These
could be useful in crosslinking cotton to make it
crease-resistant.

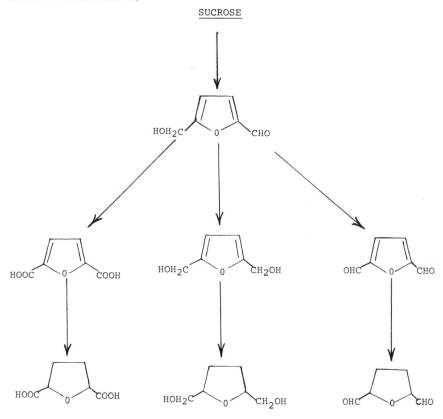

Figure 3. *Fiber precursors from sucrose (Source: Private communication from both former and present associates of Merk and Co.)*

Source: Private communication from both former and
   present associates of Merck and Company.

Let us imagine for the moment that we are talking
about a potential 30¢/lb product. Adipic acid presently lists for around 40¢/lb. Butylene glycol is around
30¢, caprolactam is around 60¢, ethylene glycol around
25¢. These are spot prices so one should treat them
with caution. However, it looks as though we are not
too far out of the ball park. Maybe one <u>could</u> make
a fiber intermediate from sugar.

As is usually the case, somebody thought of it
long before I did. As a matter of fact, there was a
substantial effort in this direction going on 20 years
ago at Merck. I talked with some of the people who
were involved in that project to see how everything
came out, because, to my knowledge, no one was making a
fiber intermediate from sugar. The first thing the
Merck group did was to find out that sugar was about
the cheapest source of carbon there was. Back in
those days it was about 2¢/lb. More important, ethylene was then selling for 4¢ and it has been a good index of petrochemical prices. So, sugar, per carbon
per pound, was about half the price. Today, as we
have seen, the ratio is reversed, but we have no guar-
antee it is going to stay that way.

Merck's first job was to make the hydroxymethyl
furfural. They got a bunch of patents on that process
which, incidentally, now are about to expire. Unfortunately they could not get the yield much over 0.4
pound/pound of sucrose. I guess that is not too surprising. We want a product with a 5-membered ring
and only half of the sugar molecule has it. That
brings us to an interesting point. Fructose was in a
different price ball park in those days than it is today. Maybe this antique is worth a second look.
Anyway Merck, even with yields up to 60%, proceeded
because price structures in those days were different
from what we now have. They made most of the products
shown on the chart, characterized them to a certain extent, even did some small scale use tests. They got
the first stage into a pilot plant and had reasonably
good success.

Merck even found some unplanned uses for some of
these materials. For example, the diol is a good selective solvent for separating aromatics from paraffins,
an area where the Udex process has been so successful.
The diol propionate exhibited fungistatic properties.
Merck even made some interesting diamines including the
one for nylon 6-6. The whole effort came to a million
and a half 1960 dollars. To duplicate this work today
would probably cost over $3 million. What happened?
"Merck decided that it had more attractive uses for its

capital than investing in the captial requirements necessary to continue this program and assuming the risks
involved in attaining large volume uses for hydroxymethyl furfural." I am not about to criticize that decision, but it does illustrate the point about coming
to a definitive decision as to the direction your business is going to take before broadening your research
efforts.

When Merck terminated things, they were not even
close to a fiber precursor. Being able to make a
monomer, or a polymer, or even a fiber from the polymer
at laboratory scale does not impress fiber people much.
They want to see enough polymer to be able to run it
through semi-commercial spinning equipment. That
means that somebody has to build not only a good sized
monomer plant, but also a good sized polymer plant
before you even know if the product is going to be
acceptable. Incidentally, there is an interesting new
high oxygen fiber out which seems to have avoided that
problem and which may give you some ideas. Cyanamid
is making polyglycolic acid into fibers, but their
fibers are being used for soluble surgical sutures.
It is a small market in tons, but a gigantic one in
dollars. Maybe Cyanamid will now turn their attention
to making disposable clothing out of polyglycolic acid.
Think of the possibilities in bathing suits!

I hope you have gained an appreciation of how difficult it is to take something from concept into commerce. It is not just a matter of sophisticated chemistry or good engineering. It is a lot of people with
different orientations working together and all going
in the same direction. Just keep in mind, if you are
a marketing person, you are not going to be able to
have much to market unless your technical people make
it for you. By the same token, if you are a technical
person, you are not going to be able to sell much unless there is somebody out there keeping a close eye
on the market. This, then, is the primary message I
would like to convey. An exercise in sugar chemistry
is just that, an exercise in research, but it is not
necessarily good business. Business is out there in
the market place. Take a look at it sometime. You
might find some fascination, you certainly will find a
lot of surprises.

Epilogue

Permit me just another couple of minutes to address
the question of why I was willing to talk about sugar
in an area where I have no expertise. The answer is

simple!   Sugar, I feel, is a bellweather of the renewable resource economy that is ahead of us.   In tropic-
al climates sugar cane grows 4 times faster than any
other cellulosic crop; twice as fast in temperate climates.   Nature used crops like this one to make all
the natural gas and oil and coal that we have.   Nature
used the whole plant.   Could we somehow figure out how
to do that?   Could we find a bug that could do it for
us?   Could we train it to do it in a time frame that
is closer to our perception of how dismal the future is,
rather than to geologic time.   If we cannot do that,
could we burn the whole sugar cane, not just the bagasse?   Could we do it in such a way that would give
back the mineral ash, especially the phosphorus, in a
way that could be readily restored to the plantation?
Have we the right sugar cane variety for this kind of
use, or should we be seeking others?   Are we adjusting
the spacing of our plants and the harvesting cycle in
the optimum way for growing sugar fuel?   Are we irri-
gating in the best way we know?   Or, should we really
go back to windmills?   We have learned a lot about
aerodynamics since the windmill was last looked at seriously.
        What has all this to do with sucrochemistry?
Despite what I said about sucrochemicals not making a
ripple in sugar markets, we must keep in mind that, no
matter what kind of activity we are in, we are going
to have to make sure we have continuity of both material and energy.   Until recently, petroleum filled both
roles, but things are changing rapidly, perhaps more
rapidly than any of us wants to realize.   It very well
may be that the economy of the future is going to have
to rely on renewable resources for both energy and materials.   I have a suspician that sugarcane, or something close to it, likely is to be the key to that future economy. Bear in mind that petrochemicals use only
5% of all the oil we pump.   It well may be that sucrochemicals will take only 5% of all the sugarcane we
grow, but I can see us growing one whale of a lot more
sugarcane than most people ever thought of.   When
that happens sugar could really become cheap, relatively.   Now, that is Long Range Planning.   And that, is
the challenge.   Do enjoy it.

Abstract

        There is great appeal today in using sugar, or any
other renewable resource, to make things.   However one
always has grave doubts about replacing an established
raw material with an untried one.   These doubts become

compelling when raw material consumption is so large as to be a significant fraction of world sugar production, or when the new raw material supplier is "strange" or when the new raw material does not "look" like the current one. This characterizes the problem.

Although sucrochemistry has fascinated generations of chemists; their accomplishments have not resulted in introducing "significant" amounts of sucrochemicals into channels of commerce. Why this is so is ascribed, in part, to failure of chemists to think in terms of Market Pull. Instead they have been addressing markets naively, if at all, while pursuing that egocentric fascination which we call Raw Materials Push. Another portion of the failure is ascribed to reluctance of sugar firms to learn about markets other than their traditional one. Efforts to move raw materials under such circumstances can be characterized as a kind of Waste Disposal Project. Rarely do they work. Illustrations of some that have and have not are presented. We will focus on how to capitalize on Market Pull thinking and on assessing advantages and disadvantages - real and perceived - that sucrose offers in specific markets.

Biographic Notes

Benjamin J. Luberoff, Ph.D., P.E., Consultant to the chem. ind. Educated at the Cooper Union (New York) and Columbia Univ. Has been a faculty member of both schools. Industrial experience includes American Cyanamid Co., The Stauffer Chemical Co., and the Lummis Co., where he was Mgr. of Process Res. He founded, in 1971, and now edits Chemtech for the Am. Chem. Soc. 48 Maple Street, Summit, N.J. 07901 U.S.A.

27

Hopes in a Sucrochemical Future

A. J. VLITOS

Tate & Lyle Ltd., Group R & D, Philip Lyle Memorial Research Laboratory,
The University, Whiteknights, P.O. Box 68, Reading, Berks, RG6 2BX, England

It is a very difficult task trying to summarize some of the papers given in the earlier sessions, and drawing together the 'Hopes in a Sucrochemical Future'. Obviously a task of this sort cannot be done lightly for it should include some of the historical perspective. I would like to begin first of all by paying tribute to the remarkable forecasting done by Dr. Hass in the early days of the Sugar Research Foundation. For, not only was it Dr. Hass who coined the term "sucrochemistry" but, in fact, he laid the ground work which led to this Symposium which we have been attending these past three days.

It must be most satisfying to Dr. Hass to have noted the broad expanse of the sucrochemistry in the papers presented in the past four sessions. But, it is important to remember that all of this work was done on a very small budget, by today's standards an extremely small budget. I am sure Dr. Hass would agree with this. Indeed it is a tribute to the early Foundation management that so much was achieved, not only with so little money, but in such a short time.

Professor Hough showed us that the potential for modifying the sucrose molecule is almost limitless. In fact, when one considers the number of items that he has described to us, I think Professor Hough could work the rest of his life synthesizing the numbers of compounds that are possible.

Not only has Professor Hough given us some interesting chemistry, he has, in fact, trained a generation of sucrose chemists, several of whom followed in the program, including, Dr. Khan, who gave a very eloquent paper on the acetates of sucrose, and Professor Hall, who showed us how n.m.r. spectroscopy can be used to identify and probe sucrochemicals. Someone, not of the Hough school, Professor Avela, gave us a very interest-

ing paper and a very good paper, on how to produce the
chelates, and use these chelates to effect direct substitutions of hydroxyls in the sucrose molecule.

One is bound to believe that the chemistry in all
of these papers provides opportunities to be exploited
in the future. This, then, was the first session
that had to do with fundamental chemistry.

In the second session, we shifted focus to the
applications of this sucrochemistry, the so-called,user
technology. Obviously, the surfactants, surface coatings, sucrose-based urethanes, all are receiving a
good deal of attention.

We heard about some new sucrose ester surfactants
from Dr. Parker of Tate & Lyle, and the stage of the
developments of older esters from Mr. Kosaka of the
Ryoto group and from Mr. Bobichon from Rhône-Poulenc.

Professor Bobalek explained how this fundamental
chemistry forms the base of the pyramid of essentials
in the exploiting of sucrochemistry by building up use-
applications. He also established some of the parameters to be considered in further development of surface coatings esters.

The production of specialty chemicals via fermentations of sucrose was shown to hold a great promise by
the several papers indicating the feasibility of such
processes for solvents, gums and protein supplements.

It was obvious from these and especially from
Dr. Weaver's and from Professor Bobalek's papers that,
because of the foundation of good research in sucrochemistry in the past, the sugar industry now is in a
much better position to enter the energy race. Through
this work, in fact, we can begin the race with a bit of
a head start. Unfortunately, there have been too many
stops and go's in the sucrochemical programs, particularly those supported by the Sugar Research Foundation,
as well as elsewhere. Because this has been true in
the past; there has emerged considerable skepticism
about sucrochemistry. The reviews by Dr. Sheppard and
Dr. Luberoff displayed some of that skepticism, and
some of it is justified. It seems important, therefore, to try to rationalize the arguments for and
against sugar as a possible substitute for petrochemicals in industrial chemistry. What antagonisms are
likely to come up? Sucrochemistry meets questions
which essentially are economic. One notes that, if
there is to be a significant increase in the amount of
cane sugar that is to be produced, it is bound to come
at a market price which is higher than the present
world price. Similarly, there will be increased pressures on sugar produced from beets. Farmers in Europe

always are demanding higher prices for their crop.
One notes, there is bound to be increased pressure
from maize, especially on the Continent. As new
maize varieties are introduced in Europe, there is
likely to be pressure for even higher beet prices.
This, in turn, will result in higher sugar prices.
     But, what is the important key, one would believe,
is in the long run differential between the increased
prices for sugar and the increased prices for petrochemicals. It seems reasonable to conclude that
petrochemical prices will increase more rapidly than
sugar prices, and that there will be enough sugar at
competitive prices to compete -- in certain, specialty
chemical markets. Now, I have stressed that, in the
short term, sucrose will compete with petrochemicals
only in specialty chemicals markets. In this we agree
with the previous speakers that we should not look for
short term, major markets for sugar that are going to
match or even approach the scale of markets in foods,
to which the industry has been accustomed. The new
opportunities are going to be in specialty markets for
high priced commodities.
     Yet, it should be stated that sucrose will not be
competing with ethylene; it will be competing with
ethylene derivatives which, in fact, cost more than
ethylene by factors of two or three.
     Now, the second question put before you concerns
investments by the oil industry, to be made in new
ethylene plants and in the American plants to exploit
coal resources. How will these affect the chances of
sucrose chemistry? The interesting point here is that
the capital investment in an ethylene plant today is
roughly four times what it was in the early '70's.
These ethylene plants are naptha cracking plants.
A recent study by a European ethylene producer indicated that he would have to sell ethylene at about £450
to £600 per ton by 1985, not even allowing for inflation, to obtain a decent return on his captial investment. So ethylene is going to be very expensive.
Whether sugar is going to be that expensive or not, can
not presently be predicted precisely but, either way,
the sugar producers would not mind very much.
     As for exploitation of coal in the U.S., this also
is bound to involve heavy capital investments. One
doubts very much that the petrochemicals which result
from a new coal exploitation will be any less expensive
than oil-derived products.
     The third question to ask is whether the chemical
industry is geared to handle carbohydrate feed stocks
as starting materials. The answer is that there are

few, if any, major chemical companies which are seriously developing alternative chemical feed stocks.
One or two smaller companies have expressed interest in sucrose chemistry, but these mainly, are for specialty chemical applications. One may ask how many sugar companies have considered seriously diversifying into the chemical field. Either alone or in partnership with a chemical company, are they willing to develop those sucrose chemicals which seem economically viable today and likely to become even more so in the future?

One may conclude that there are two potential weaknesses in the present sucrochemistry programs. The program is unlikely to convince the giants in the chemical industry to adopt, what is for them, a new technology, as well as an unfamiliar feedstock. The sugar industry, itself, seems at the moment to lack the expertise or the entrepreneurship to enter the chemicals markets. This may sound rather pessimistic, but perhaps it is realistic.

Those in the sugar industry will have a possibility if they can find some effective means of effecting cooperation with the chemical industry, perhaps in joint ventures of the type displayed in the Ryoto Company, organized between Dai Nippon Sugar and Mitsubishi Chemicals. It is this sort of link and joint venture, that seems to be the way most likely really to develop and exploit the fundamental sucrochemistry for which the International Sugar Research Foundation has been paying for this third of a century.

However, it would be a major mistake to overlook the potentials of sucrochemistry just because its commercial viability has yet to be realized. One must keep in mind there was a time when petrochemical's or petroleum's major markets were in patent medicine and cough drops. One can be convinced that the sugar chemistry, the subject of this Symposium for the past few days definitely will lead to some of the specialty products of the late '80s or early '90s.

On this optimistic note, I would like to remind you of Oscar Wilde's apt phrase that, "It's dangerous to prophesy, especially about the future".

Abstract

It is timely to be thinking about a "sucrochemical future" at a time when there is considerable concern about shortages of conventional sources of energy. For sucrose represents a unique substance - one of the major products resulting from the conversion of solar

to chemical energy by higher plants.   This fact is
going to be of increasing significance because, in the
long run, it is the products of photosynthesis which
may have to replace our diminishing supplies of fossil
fuels.   Thus,when one looks into the "sucrochemical
future" it is to anticipate the use of sucrose and
other sugars as chemical feedstocks or as fermentation
substrates to produce a vast array of chemical products
which today depend almost entirely upon petrochemical
feedstocks.

This paper discusses those processes based on sucrochemistry which stand the best chance of competing
economically with petrochemicals both in the short- and
long-term.   Special emphasis is placed on the use of
sucrose to produce sucrose surfactants employing simple
techniques or, at the most, intermediate level technologies.

Biographic Notes

Professor A.J. Vlitos, Ph.D., Dir. and Chief Exec.
of R & D.   Educated at Oklahoma State Univ., Harvard
Univ., Iowa State Coll. and Columbia Univ.   He was a
senior plant physiologist at the Boyce Thompson Inst.
for Plant Res. Inc.; in 1959 he joined Tate & Lyle Ltd.
in Trinidad, and in 1966 was appointed to his present
post.   He is Honorary Visiting Prof., Depts.of Biology
and Chem., Queen Elizabeth College and Visiting Prof.
at Univ. of Reading.   Philip Lyle Memorial Research
Laboratory, Univ. of Reading (Tate and Lyle, Ltd.).
P.O. Box 68, Berkshire RG6 2BA, England.

Discussion

Question: Dr. Marcy, how do the ISRF patenting
and licensing programs and the license terms used fit
into the other programs of patenting and licensing that
Research Corporation carries on?   Are the licensing
terms typical, or are they unusual?   How do you compare the ISRF program with those carried on for other
people?

Dr. Marcy: It all depends on what kind of technology you are talking about.   The general terms in
the ISRF-type of license are very similar to the general terms in any other licenses.   As I mentioned
in my talk, there is no basic difference between licensing sucrochemicals and other types of chemicals.
This leads one to compare, then, the fine structure
of a license.   If one is talking about a pharmaceutical, or a drug, or some fine chemical, that has a relatively limited market, generally the royalty rates
are higher; a common arrangement is 5 to 10 percent of
the net sales price of the product in the final pack-
aged form.   Royalties are collected on the package as
well as the material.   With sugar esters, and intermediate and heavy chemicals, normally this is not possible.   Here, the royalty rate is based on the net
sales value of the product in bulk form to wholesalers.
These royalty rates may run from 3 to 7 percent.
I think that the original figure for sugar ester
licenses, two percent, was on the low side; it could
have been up to five percent.   Nonetheless, the
Japanese licensee came back after a few years and said
that they thought two percent was too high, and so the
royalty rates were reduced, using a sliding scale down-
ward, based on the volume of products.   That is another usual way of doing it.
In the electronics and electrical industry, gener-

ally royalty rates run lower.   They run from 0.5 percent up to, perhaps, three percent.   In the mechanical
arts area, license rates will be in that same range.
It is more difficult to license mechanical devices on
a percentage of selling price basis, as mechanical devices frequently are components of a larger piece of
machinery.   In this situation it is preferable to set
the royalty on a cents per unit basis.   I do not think
that is applicable in licensing chemicals.

Question: Dr. Marcy, you mentioned that you have
four ISRF patent applications in process at the moment.
Would you tell what they are?

Dr. Marcy: With permission from ISRF I would be
pleased to.

Mr. Sarault: You have my permission.

Dr. Marcy: One covers the inventions and discoveries of Mr. Faulkner, of the Paint Research Association,
who described his work on surface coating sucrose resins (Chapter 13).   A second covers the discoveries and
inventions of Dr. Poller in the metallo-sucrose derivative area (Chapter 11).   A third covers inventions and
discoveries of Dr. Gardner, whose work has not been
discussed at this symposium.   His technology involves
a sucrose-based polymeric material for use in dimen-
sionally stabilizing wood.   The fourth one covers
Professor Hough's work on chlorosucroses (Chapter 2).
Applications covering all of these inventions are being filed in the United Kingdom first and then, within
the "convention year," in the United States and in
perhaps 15 to 20 countries around the world.

Question: What is the approximate world use of
sucrose esters?

Dr. Marcy: I do not have that information available.   As I mentioned, there are two major licensees,
one in Japan and one in France.   The sales of these
materials can be approximated from the reports of the
royalty payments, which, again only with permission
from ISRF may I disclose.

Mr. Sarault: You have my permission.

Dr. Marcy: Last year's royalty payment was $35 -
$40,000.   At a two percent royalty rate, sales of
about $2,000,000 would be indicated for these two coun-

tries.   The sucrose ester patents expired this year, so this year's royalties are not really indicative. In addition, Tate & Lyle is marketing an ester in the United Kingdom, and there are small amounts of sugar esters manufactured by other processes that are market- ed in some other countries, particularly in Belgium, Holland and Germany.   But, these are minor.   As far as I know, there are no sales in Italy, the United States and Canada.   In other countries of the world, as far as I know, there are no sales.

Someone more familiar with the actual marketing of esters may have some additional information.   Do you, Professor Vlitos?

Professor Vlitos: No, I do not have any figures.

Dr. Marcy: Does anyone know if there is any manu- facture and sale of sugar esters in the Eastern bloc countries?

Dr. Hickson: The Russians have some patents on sugar esters.

Dr. Marcy: Patents are not equivalent to sales, unfortunately.

Question: What is the cost of worldwide patent coverage, and does the cost go up over the years? And, is it an economical thing to do, unless you are absolutely certain that the products have real commer- cial possibilities?

Dr. Marcy: One is never absolutely certain about anything.   Therefore, in this area, as with any chemi- cal research, you have to make some value judgments. These materials and this work that has been done with ISRF money are still quite basic.   Much additional funding has to be expended for further research and development.   If you add to this the money that one would have to spend to get Food and Drug Administration clearances on some of these products, you reach into the millions of dollars before you even get on the market.   Value judgments on whether or not to spend money for patenting can be affected by these factors. My personal feeling is that the cost of patenting is relatively small compared to the cost of further re- search and development and market research.   Generally speaking, people tend to file for patents and prosecute them fairly quickly, with very little assurance that they are going to have any commercial products.

You also had a question asking if patent coverage costs more as time goes on. Practically all of the countries of the world that have patent systems have annual maintenance fees, or working requirements for keeping patents in force. In some countries, the fees are progressively greater each year during the life of the patent, and sometimes they escalate very rapidly. Therefore, what one does is to file first, at a relatively low cost, and then try to get the product on market as soon as possible. If the products are valuable commercially, they will pay for the added costs of maintenance. If they are not being used commercially in a given country, simply do not pay the maintenance fees, and the patent becomes abandoned. In the United States and Canada there are no maintenance fees, although they have been discussed. Thus, in these countries, the total costs are simply the original filing and prosecution costs.

A U.S. patent application of the sort that one would file in the case of sucrochemicals probably would cost in the order of $2,000 to $3,000, just for the filing. This charge covers the attorney who drafts and files the application and the filing fees. Filing in other countries, once filing has been accomplished in one country, is simpler and less costly per country, except for translation costs. Translation costs are fairly high in Japan, but in a French-speaking or German-speaking country, the cost is on the order of a few hundred dollars per patent. Filing in 15 major couutries around the world would cost from $10,000 to $15,000.

Question: Would you tell us, roughly, what is the breakdown of the $10,000 to $15,000; what fraction does Research Corporation pay and what fraction devolves upon ISRF?

Dr. Marcy: The agreement between Research Corporation and ISRF calls for Research Corporation to evaluate inventions that are submitted to Research Corporation by ISRF, and make a decision as to whether it is worthwhile to go forward with patenting and licensing. If we decide in a positive direction, then we undertake to get patent coverage,and maintain those patents; to handle any interferences, or infringements that might come up during the life of the patent; and to collect the royalties, all at our expense. ISRF is charged nothing for this service. If and when income is received, however, the agreement calls for Research Corporation to share it with the inventor (at whatever percentage the inventor is designated to get, by ISRF) and ISRF. After the payment to the inventor, the distribution between the two organizations is 50:50.

Dr. Weaver: By the time these proceedings are published, Proctor and Gamble will have disclosed that the sucrose octaester of a higher fatty acid, such as stearic, will inhibit the deposition of cholesterol. It boggles the mind what might happen to sugar ester production figures if even one gram of octaester of sucrose were put into each pint of Crisco.

INDEX

INDEX